MEMBRANE AN

The INTRODUCTION TO BIOTECHNIQUES series

CENTRIFUGATION
RADIOISOTOPES
LIGHT MICROSCOPY
ANIMAL CELL CULTURE
GEL ELECTROPHORESIS: PROTEINS
PCR, SECOND EDITION
MICROBIAL CULTURE
ANTIBODY TECHNOLOGY
GENE TECHNOLOGY
LIPID ANALYSIS
GEL ELECTROPHORESIS: NUCLEIC ACIDS
LIGHT SPECTROSCOPY
DNA SEQUENCING
MEMBRANE ANALYSIS

Forthcoming title

PLANT CELL CULTURE

MEMBRANE ANALYSIS

John M. Graham

John Graham Research Consultancy, Peter Jost Enterprise
Centre, Liverpool John Moores University, Byrom Street,
Liverpool L3 3AF, UK

Joan A. Higgins

The University of Sheffield, Department of Molecular Biology
and Biotechnology, P.O. Box 594, Firth Court, Western Bank,
Sheffield S10 2TN, UK

Taylor & Francis
Taylor & Francis Group

LONDON AND NEW YORK

© Taylor & Francis, 1997

First published 1997

Transferred to Digital Printing 2006

A CIP catalogue record for this book is available from the British Library.

ISBN 1 87274 888 0

Published by Taylor & Francis
2 Park Square, Milton Park, Abingdon, Oxon, OX14 4RN
270 Madison Ave, New York NY 10016

Production Editor: Els Boonen
Typeset by Chandos Electronic Publishing, Stanton Harcourt, UK.

Publisher's Note
The publisher has gone to great lengths to ensure the quality of this reprint but points out that some imperfections in the original may be apparent

Contents

Abbreviations

BCA	bicinchoninic acid
BSA	bovine serum albumin
CAT	chloramphenicol acytltransferase
CRD	cross-reacting determinant
DPG	diphosphatidylglycerol
ELISA	enzyme linked immunoassay
ER	endoplasmic reticulum
FAMEs	fatty acid methyl esters
GAG	glycosaminoglycan
GLC	gas-liquid chromatography
HPLC	high performance liquid chromatography
HPTLC	high performance thin-layer chromatography
IEF	isoelectric focussing
LMP	light mitochondrial fraction
LUVs	large unicellular vesicles
MAb	monoclonal antibody
MLVs	multilamellar vesicles
NANS	N-acetylneuraminic acid
PC	phosphatidylcholine
PE	phosphatidylethanolamine
PG	phosphatidylglycerol
PS	phosphatidylserine
PT	phosphatidylinositol
RER	rough endoplasmic reticulum
SDS-PAGE	sodium dodecyl sulfate-polyacrylamide gel electrophoresis
SER	smooth endoplasmic reticulum
SM	sphigomyelin
SUVs	small unicellular vesicles
TLC	thin layer chromatography

Preface

A brief examination of the program from any international or national congress in biochemistry, cell biology, molecular biology or immunology, reveals that subcellular membrane isolation and analysis form major elements in a wide range of research interests. Secretion and internalization of physiologically important molecules, ion transport, energy transduction, intracellular and intercellular signalling and the biogenesis of cell components are just a few of the areas of intensive research activity which involve membrane analysis. These exciting areas are underpinned by the continuing and crucial basic research on how membranes are organized and function.

Unlike most other books in the Introduction to Biotechniques series, this text is organized around a broadly based research topic rather than a more defined technical one; therefore the first two chapters provide some background information on membrane composition and membrane types, particularly as these relate to later chapters on fractionation and identification (Chapter 3) and on compositional and structural analysis (Chapter 4). Techniques in these two chapters are prodigious and in an introductory text they cannot be reviewed in detail. The authors' approach to this problem is either to provide a few sample protocols, or to describe only the principles of individual techniques and their applicability and to highlight important practical points. Many of these techniques are subsequently used as part of the methodologies described in Chapters 5–8.

These subsequent chapters on membrane structural/functional relationships have necessarily been rather selective and their choice has been partly influenced, by the authors' own research interests. Nevertheless, the topics covered in Chapters 5–8, protein and lipid topography, protein targetting, membrane trafficking and membrane biogenesis, should be of relevance to a broad spectrum of research workers. Here too, the emphasis is on technical principles, as detailed protocols often have to be tailored to specific situations and sample types and dictated by operational requirements.

This book is aimed primarily at the new research worker, but we hope that it will also be of considerable benefit to more experienced workers who will find the more basic information in Chapters 1–4 a useful source of revision, and the new and advanced technologies described in Chapters 5–8 a valuable aid to their research. In addition to the

Table 1.1. Protein:lipid ratios in membranes

Membrane	Protein:lipid	Membrane	Protein:lipid
Animal			
Myelin	0.25	Erythrocyte	1.1
Surface (liver)	1.5	Nuclear (liver)	2.0
Rough ER (liver)	2.5	Smooth ER (liver)	2.1
Inner mit. (liver)	3.6	Outer mit. (liver)	1.2
Golgi (liver)	2.4	Sarcoplasmic reticulum	3.0
Retinal rod	1.0		
Plant			
Surface	0.9	Chloroplast	1.9
Saccharomyces			
Surface	1.2		
Prokaryote surface			
Bacillus	2.8	*Micrococcus*	2.4
Staphylococcus	2.4	*Escherichia coli*	2.8
Outer membrane of	2.2		
Gram-negative bacteria			

> Because protein is much denser than lipid, membranes with different protein:lipid ratios may also have different densities – one of the properties that is used to fractionate membranes by centrifugation (see Chapter 3).

Generally lipids can be divided into four major classes:

* free fatty acids;
* esters of fatty acids, e.g. triacylglycerol and phospholipids;
* isoprenoids, e.g. sterols, sterol esters, dolichol and farnesol;
* glycolipids.

Phospholipids are major components of all membranes and, with the exception of myelin, they represent greater than 50% of the total membrane lipid mass. Glycolipids and cholesterol are usually concentrated in the plasma membrane of mammalian cells, while in plants only sterols (mostly stigmasterol and sitosterol) show this preferential localization and glycolipids predominate in the chloroplast. Bacterial membranes may also be rich in glycolipid but, with the exception of mycoplasmas, they lack sterols, and even in mycoplasmas the sterol is derived almost exclusively from the growth medium. It is only relatively recently that isoprenoids other than sterols (dolichol phosphate and derivatives of farnesol and geranyl-geranol) have been shown to be important in membranes. The relative amounts of glycolipid, phospholipid and cholesterol in some examples of membrane types is given in *Table 1.2*.

Table 1.2. Lipid composition of membranes (figures are % of total lipid mass)

Membrane	Phospholipid	Glycolipid	Sterol
Mammalian			
Plasma	50–60	5–17	15–22
Endoplasmic reticulum	70–80	<5	5–10
Mitochondria (inner)	80–90	<5	<5
Mitochondria (outer)	80–90	<5	5–8
Lysosomes	70–80	5–10	10–15
Nuclear	85–90	<5	10–15
Golgi	85–90	<5	5–10
Peroxisomes	90–95	<5	<5
Myelin	50–60	15–25	20–25
Erythrocyte	70–80	5–10	20–25
Plant			
Surface	30–65	10–20	25–50
Mitochondria	90–95	<5	<5
Chloroplast envelope	20–30	65–80	<5
Chloroplast lamellae	35–45	50–70	<5
Endoplasmic reticulum	70–80	5–15	10–20
Bacterial			
Cytoplasmic	50–90	10–50	None

Nonpolar lipids such as cholesterol ester, and triacylglycerol are rarely encountered as membrane components; if they are present in an isolated membrane fraction it is normally within a membrane vesicle as part of a lipoprotein, for example, which is destined for secretion from the cell. The endoplasmic reticulum (ER) which is the site of synthesis of triacylglycerol and cholesterol ester may, however, contain small amounts of these lipids. Free fatty acids are also not considered as discrete membrane components, although they may arise from the degradation of phospholipids, either as part of a biological process, or artificially during isolation of membranes (see Chapter 3) as a result of exposure of the lipids to hydrolytic enzymes which are normally sequestered *in vivo*. Like the farnesyl and geranyl-geranyl isoprenoid residues, however, some fatty acids (e.g. palmitic or myristic acid) may anchor proteins to the lipid bilayer (see Sections 1.1.3, 1.2.3, 5.2.3 and 7.2.11).

Degradation of membrane lipids during fractionation must be limited by careful control of temperature (4°C) and the general avoidance of Ca-containing media which can activate some phospholipases. Free fatty acids cause uncoupling of phosphorylation in mitochondria.

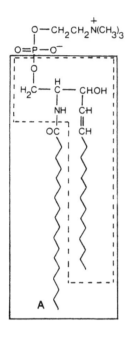

O—CH$_2$CH$_2$N(CH$_3$)$_3$$^+$
O=P—O$^-$

H$_2$C—C—CHOH
| |
NH   CH
| ||
OC   CH

A

O
|
O-P-O-CH$_2$
| CH
O |
CH$_2$
O

Ethanolamine
or choline

B

Figure 1.2. Molecular structure of
sphingomyelin (A) and ether lipid
(B). In A, structure in dotted box is
sphingosine, that in continuous box
is ceramide. Reproduced from *Cell
Biology Labfax*, Dealtry, G.B. and
Rickwood, D. (eds) (1992) with
permission from BIOS Scientific
Publishers.

Branched chain 'iso' fatty acids

$$CH_3$$
$$CH_3-CH-(CH_2)_n-COOH$$

Branched chain 'anteiso' fatty acids

$$CH_3$$
$$CH_3-CH_2-CH-(CH_2)_n-COOH$$

Cyclopropane fatty acids

$$CH_2$$
$$CH_3-(CH_2)_5-CH-\!\!\!-\!\!\!-CH-(CH_2)_9-COOH$$

Figure 1.3. Molecular structure of some typical fatty acids in bacterial
phospholipids. Reproduced from *Cell Biology Labfax*, Dealtry, G.B. and
Rickwood, D. (eds) (1992) with permission from BIOS Scientific Publishers.

Table 1.3. Major fatty acids of eukaryotic phospholipids

Carbon number	Number of double bonds	Name	Formula
12	0	Laurate	$CH_3-(CH_2)_{10}-COO^-$
14	0	Myristate	$CH_3-(CH_2)_{12}-COO^-$
16	0	Palmitate	$CH_3-(CH_2)_{14}-COO^-$
16	1	Palmitoleate	$CH_3-(CH_2)_5-CH=CH-(CH_2)_7-COO^-$
18	0	Stearate	$CH_3-(CH_2)_{16}-COO^-$
18	1	Oleate	$CH_3-(CH_2)_7-CH=CH-(CH_2)_7-COO^-$
18	2	Linoleate	$CH_3-(CH_2)_4-(CH=CH-CH_2)_2-(CH_2)_6-COO^-$
18	3	Linolenate	$CH_3-CH_2-(CH=CH-CH_2)_3-(CH_2)_6-COO^-$
20	0	Arachidate	$CH_3-(CH_2)_{18}-COO^-$
20	4	Arachidonate	$CH_3-(CH_2)_4-(CH=CH-CH_2)_4-(CH_2)_2-COO^-$
22	0	Behenate	$CH_3-(CH_2)_{20}-COO^-$
24	0	Lignocerate	$CH_3-(CH_2)_{22}-COO^-$

generally absent from prokaryotes. In this brief review of membrane structure, however, we cannot do justice to the huge diversity of fatty acid components in bacteria, which in many cases reflects the growth conditions of the organism.

Acyl residues in mammalian membranes show some degree of specificity for both phospholipid (*Table 1.4*) and membrane type. The longest acyl chains (C24) are confined to sphingomyelin and are therefore prominent in myelin. Acyl residues in plant cells are highly membrane specific.

Table 1.4. Fatty acid composition of eukaryotic plasma membrane phospholipids [figures for each fatty acid are a % of the total fatty acid for each phospholipid in the rat liver plasma membrane or for the human erythrocyte membrane (figures in brackets)]

Fatty acid	% of total				
	PC	PE	PI	PS	SM
16:0	32(37)	30(23)	30	(8)	36(35)
16:1	3(2)	1(2)	8	(2)	4(2)
18:0	35(13)	31(13)	36	(37)	25(13)
18:1	10(23)	10(15)	13	(14)	2(3)
18:2	8(16)	6(7)	2	(3)	1(3)
20:0	–	–	–	–	–
20:3	–	1(1)	–	–	–
20:4	8(8)	16(23)	8	(23)	18(–)
22:6	1(–)	3(2)	–	(2)	–
24:0	–	–(2)	–	(2)	–(20)
24:1	–	–(2)	–	(2)	–(15)

1.1.3 Isoprenoids

Sterols. The amounts of cholesterol in mammalian membranes vary considerably: the mole ratio of cholesterol:phospholipid is highest in the plasma membrane (approx. 0.8 in the plasma membrane of hepatocytes and nearer 1.0 in myelin) and much lower in the cytoplasmic membranes, the inner mitochondrial membrane being essentially devoid of this lipid, while other membranes such as the endoplasmic reticulum have ratios of about 0.3. Cholesterol is almost invariably present in membranes as the free sterol, not as the ester, which is not amphipathic and does not fit easily into the lipid bilayer. Plant membranes contain either stigmasterol or sitosterol rather than cholesterol (*Figure 1.4*).

Linear isoprenoids. Dolichol phosphate (*Figure 1.5*) is present chiefly in the endoplasmic reticulum where its phosphate group is orientated to the cisternal space and it is responsible for the transfer of a branched oligosaccharide to an asparagine residue of a nascent glycoprotein. It is this mannose-containing oligosaccharide that is the starting point for the synthesis of N-linked glycoproteins. The farnesyl and geranyl-geranyl isoprenoids (*Figure 1.5*) are covalently attached to some membrane proteins, anchoring them to the lipid bilayer. It is becoming increasingly apparent that these lipid-modified proteins play important roles in cell signaling and in the regulation of cellular events (see Chapters 5 and 7).

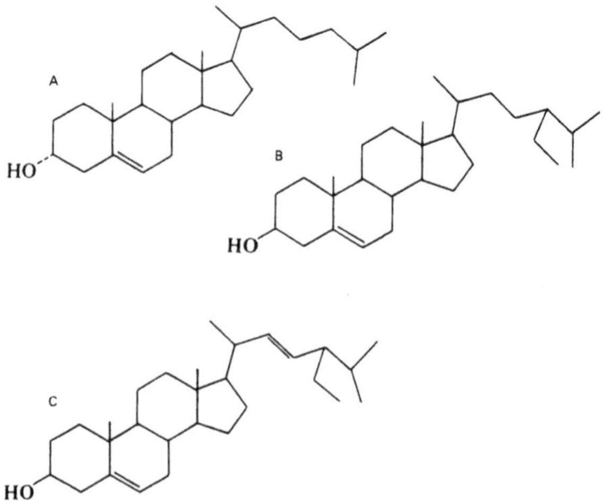

Figure 1.4. Molecular structure of sterols. A, cholesterol; B, sitosterol; C, stigmasterol. Reproduced from *Cell Biology Labfax*, Dealtry, G.B. and Rickwood, D. (eds) (1992) with permission from BIOS Scientific Publishers.

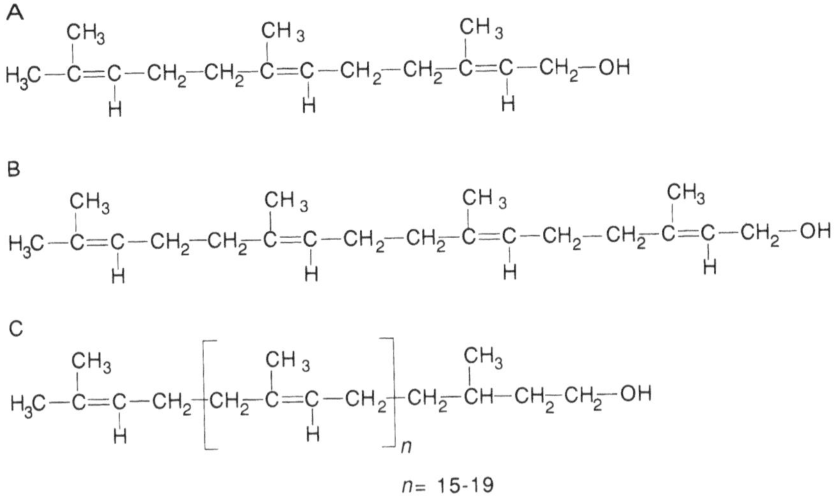

A

$H_3C-\overset{\underset{\displaystyle |}{CH_3}}{C}=\overset{\underset{\displaystyle |}{H}}{C}-CH_2-CH_2-\overset{\underset{\displaystyle |}{CH_3}}{C}=\overset{\underset{\displaystyle |}{H}}{C}-CH_2-CH_2-\overset{\underset{\displaystyle |}{CH_3}}{C}=\overset{\underset{\displaystyle |}{H}}{C}-CH_2-OH$

B

$H_3C-\overset{\underset{\displaystyle |}{CH_3}}{C}=\overset{\underset{\displaystyle |}{H}}{C}-CH_2-CH_2-\overset{\underset{\displaystyle |}{CH_3}}{C}=\overset{\underset{\displaystyle |}{H}}{C}-CH_2-CH_2-\overset{\underset{\displaystyle |}{CH_3}}{C}=\overset{\underset{\displaystyle |}{H}}{C}-CH_2-CH_2-\overset{\underset{\displaystyle |}{CH_3}}{C}=\overset{\underset{\displaystyle |}{H}}{C}-CH_2-OH$

C

$H_3C-\overset{\underset{\displaystyle |}{CH_3}}{C}=\overset{\underset{\displaystyle |}{H}}{C}-CH_2-\left[CH_2-\overset{\underset{\displaystyle |}{CH_3}}{C}=\overset{\underset{\displaystyle |}{H}}{C}-CH_2\right]_n CH_2-\overset{CH_3}{\underset{|}{CH}}-CH_2-CH_2-OH$

$n= 15\text{-}19$

Figure 1.5. Molecular structure of linear isoprenoids. A, farnesol; B, geranyl-geranol; C, dolichol.

1.1.4 Lipid architecture

As shown in *Figure 1.6*, all membrane lipids are arranged as a bilayer with the water-soluble portions of the molecules (phosphate groups, nitrogenous bases, glycerol, inositol, hydroxyl groups and

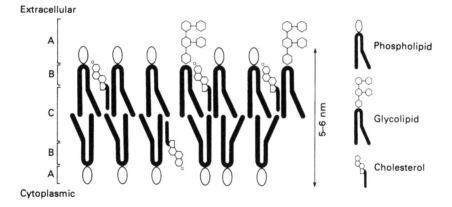

Figure 1.6. Disposition of major lipids in the bilayer of a mammalian plasma membrane. A, Polar regions; B, nonpolar regions in which the fluidity is modulated by the presence of sterol and acyl chain unsaturation; C, fluid nonpolar region. Reproduced from *Membrane Structure and Function,* Evans, W.H. and Graham, J.M. (1989) with permission from IRL Press at Oxford University Press.

carbohydrate residues) on the outside and hydrocarbon chains in the core. In the same manner, the planar ring system of the sterol (if present) and the aliphatic chains of dolichol are within the nonpolar region with the free hydroxyl or phosphate group at the surface. Acyl, farnesyl and geranyl-geranyl residues attached to proteins are similarly located in the bilayer core.

Membrane phospholipids exhibit transverse asymmetry, so that the composition of the two leaflets of the bilayer are different. Studies on the human erythrocyte membrane (see *Table 1.5*) and on the plasma membrane of a number of cell types, as well as some cytoplasmic membranes from rat liver, suggest that this asymmetry is a characterstic of all biological membranes. For more information on the generation and maintenance of lipid asymmetry see Section 6.4.

There are essentially two nonpolar regions in the lipid bilayer: a central fluid nonpolar region and a more peripheral nonpolar region where the fluidity is modulated by the degree of unsaturation of the acyl chains and by the presence of sterol (*Figure 1.6*). The fluidity of this region increases with the degree of unsaturation of the acyl chains and is reduced by the presence of a sterol. As the acyl chain becomes more unsaturated, so the transition temperature between cystalline and fluid states is reduced. The result is that under physiological conditions membrane lipids are in a fluid state, allowing lipids to diffuse laterally in the plane of the membrane. Fluidity is very important, as it allows for the rapid deformation of the membrane which occurs during fusion and endocytosis (see Chapters 7 and 8).

1.2 Proteins

Membrane proteins were classified in the Singer–Nicholson model into one of two categories based upon the method needed to remove them from the membrane.

Table 1.5. Lipid asymmetry of the lipid leaflets of the human erythrocyte membrane (figures are % of each phospholipid in each leaflet)

Lipid type	Inside	Outside
Sphingomyelin	17	83
Phosphatidylcholine	26	74
Phosphatidylethanolamine	77	23
Phosphatidylserine	95	<5
Glycolipids	<5	95

- *Peripheral proteins* can be removed by modulation of pH or ionic strength, while the basic membrane bilayer structure is maintained.
- *Integral proteins* span the membrane and can only be removed by disrupting the membrane structure with detergents.

The discovery of proteins linked to the membrane by a lipid (acyl, farnesyl or geranyl-geranyl) anchor has made this distinction less clear (*Figure 1.7*). While the polypeptide chain of such proteins is peripheral it cannot be removed by changing the pH or ionic strength of the medium; the link to its lipid anchor can, however, be cleaved enzymically thus releasing the protein without disrupting the membrane. For the purposes of this book, these proteins are considered as a third group: *lipid-linked proteins*.

Proteins provide:

- selective ionic permeability;
- the means of energy transduction;
- the means of responding to a signal on one side and of propagating a response on the other side;
- transport systems for hydrophilic metabolites (e.g. glucose and amino acids);
- structural functions through interactions with nonmembrane macromolecules in the cytoskeleton and intercellular matrix.

1.2.1 Peripheral proteins

Peripheral proteins are typically globular proteins bound ionically to the polar domain of another protein or possibly to the polar head groups of phospholipids (*Figure 1.7*). These proteins provide a structural or functional link to an integral protein or between integral proteins; or between an integral protein and a protein of the cytoskeleton or extracellular matrix.

Cytochrome c is a typical example of this type of protein which is bound peripherally to the inner mitochondrial membrane. It is the only cytochrome which is not part of a respiratory complex buried within the membrane and it serves as an electron shuttle between Complex III and cytochrome oxidase. The F_1 subunit of ATP-synthase is another example.

Peripheral proteins on the outside of membranes are prone to removal during homogenization and fractionation procedures.

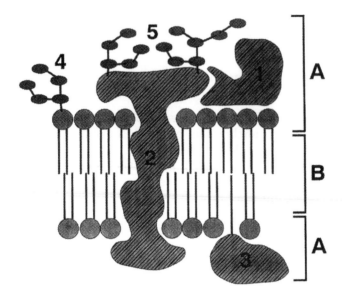

Figure 1.7. Disposition of major protein types in lipid bilayer of the plasma membrane. A, Polar regions (extracellular space at top, intracellular space at bottom); B, nonpolar regions; 1, peripheral protein bound ionically to phopholipid head groups and/or to the extracellular domain of: 2, an integral glycoprotein; 3, a lipid-anchored protein; 4, oligosaccharide chains of a glycolipid; 5, oligosaccharide chains of the glycoprotein.

1.2.2 Integral proteins

In the simplest of these proteins, the polypeptide chain passes once completely across the lipid bilayer and contains three principal domains (*Figure 1.7*). A median amino acid sequence (22–25 residues), rich in hydrophobic residues and normally (though not exclusively) α-helical in conformation, spans the nonpolar core of the membrane. This domain separates the extracellular and cytoplasmic domains which are rich in polar amino acids. The two polar domains may contain as many as 500–600 amino acid residues, although if the cytoplasmic domain is very large the extracellular domain is normally relatively small and vice versa. The N-terminus is normally on the outside, although the reverse is occasionally observed (e.g. transferrin).

In many integral proteins, the extracellular and cytoplasmic domains are separated by several (up to 14) hydrophobic sequences which span the membrane and which are joined by short (usually) loops containing polar residues (see Chapter 5).

As the amino acid sequences of more and more integral proteins have become established, they have been grouped into families or superfamilies on the basis of shared structural motifs. For example, in certain integral proteins amino acids in the cytoplasmic domain immediately adjacent to the transmembrane domain are often highly basic and in some of the proteins involved in phosphorylation, serine and threonine residues are common close to the C-terminus. These are just two of the many motifs shared by certain families of integral proteins.

Integral membrane proteins may exist as oligomeric units in the plasma membrane. Transferrin for example comprises two identical subunits linked by a single S–S bridge; the insulin receptor too is made of two identical units each of which comprises an α-subunit (essentially a peripheral protein) linked by an S–S bridge to the transmembrane β-subunit. The nicotinic acetylcholine receptor is more complex consisting of five subunits. The amino acid sequence of the α-helical domain of oligomeric proteins may be predominantly hydrophobic on one side (adjacent to the lipid bilayer) but predominantly hydrophilic on the side facing the other helices, thus constructing a polar channel which can be effectively 'opened or closed' by conformational changes in the subunits upon binding of the appropriate effector.

Membrane protein topography and targetting sequences are considered in Chapters 5 and 7.

Integral proteins are important as:

- receptors which are involved in the internalization of a ligand;
- receptors linked to cytoplasmic signaling;
- receptors that are ion channels;
- some antigen receptors;
- communication links between extracellular and intracellular proteins, particularly between proteins of the cell coat and the cytoskeleton;
- transporters of ions or small water-soluble molecules (e.g. sugars);
- energy transducers.

The existence of specific protein domains on integral membrane proteins has important consequences for the raising of antibodies for the isolation and identification of both surface and intracellular membranes.

1.2.3 Lipid-anchored proteins

The lipid may be a unique form of glycolipid comprising phosphatidylinositol covalently linked to a carboxy side chain on the protein by an ethanolamine residue and a branched chain of eight sugar residues. Other lipid-anchored proteins have a more simple anchor – a covalently bound fatty acid – often myristic acid linked via an amide bond to an N-terminal glycine. See Chapter 5 for structural details. Palmitoylation of a cysteine residue can achieve a similar result, although with the *ras* oncogene proteins palmitoylation appears to be dependent on prior linking of a farnesyl or geranyl-geranyl residue. Many of the so-called ectoenzymes on the plasma membrane, such as 5′-nucleotidase and alkaline phosphatase, are apparently lipid-linked, but perhaps the most important group of lipid-linked proteins are the G-proteins which play such an integral part in many transmembrane events (see Chapter 7).

1.2.4 Protein architecture in the plane of the membrane

Like lipids, many proteins are able to diffuse in the plane of the membrane, although 100–100 000 times more slowly than the lipids. Lateral diffusion of protein molecules may be an important part of processes such as endocytosis in which receptors (once a ligand has bound) must migrate to, for example, a coated pit. However, there are membrane features which may restrict lateral mobility of protein molecules, such as tight junctions and the interaction with membrane-associated structures such as the cytoskeleton and glycocalyx which may create specialized domains and functional polarity in certain cell types (see Section 2.3.1).

1.3 Carbohydrates

Carbohydrates are present as short oligosaccharide chains (often branched) and bound covalently to protein or lipid molecules as glycoproteins and glycolipids. In eukaryotes, glycolipids and glycoproteins are:

- highly site-specific; although they are present in the cytoplasmic membranes of the cell during synthesis, the mature forms are most highly concentrated on the plasma membrane;
- invariably arranged with their oligosaccharide chains on the extracellular surface of the plasma membrane, in accordance with their major roles as surface receptors and antigens.

Proteoglycans only occur in eukaryotes. They consist predominantly of carbohydrate, with long (largely unbranched) chains of sugar derivatives, called glycosaminoglycans (GAGs), linked to relatively short polypeptide segments. Proteoglycans are confined exclusively to the outer surface of the plasma membrane where they form the gel-like mesh of the extracellular matrix. Some proteoglycan molecules retain a link to the plasma membrane via a terminal hydrophobic amino acid sequence, but in many cases this part of the polypeptide chain is cleaved off. Proteoglycans are involved in cell–cell adhesions; communication between cells and their flexible GAG chains can provide an anchor to some enzymes whose activity resides at the cell–plasma interface (e.g. lipoprotein lipase).

Some of the commonly used abbreviations for sugars and their derivatives are given in *Table 1.6.*

1.3.1 Glycolipids

Glycolipids have the same overall form as phospholipids, i.e. they are amphipathic and the hydrophobic part is made up of two hydrocarbon chains. The polar part, however, as the name implies, is made up of a series of monosaccharide residues. Like phospholipids, glycolipids are also based on either diacylglycerol or N-acylsphingosine.

Glycosylated derivatives of diacylglycerol. These glycolipids (*Figure 1.8*) are found in the membranes of bacteria and plants, but very rarely in animal membranes: the 3-OH group of diacylglycerol, which is linked to a phosphate-base in phospholipids, is replaced by a short oligosaccharide (2–4 residues). Mono- and digalactosyl diacyl-glycerols are the most common in plants. In bacteria the glycosyl groups can be more complex and include glycosyl derivatives of phosphatidate (e.g. phosphatidyldiglucosyldiglyceride) and of PI in which one or more of the inositol OH groups are mannosylated.

Table 1.6. Abbreviations for some of the sugars and sugar derivatives in glycoproteins and proteoglycans

Sugar	Abbreviation	Sugar	Abbreviation
Glucose	Glc	Fucose	Fuc
Galactose	Gal	D-Mannose	Man
N-acetyl-D-glucosamine	GlcNAc	D-Xylose	Xyl
N-acetyl-D-galactosamine	GalNAc	D-Glucuronic acid	GlcUA
N-acetylneuraminic acid	Neu5Ac or NANA	L-Iduronic acid	IdUA

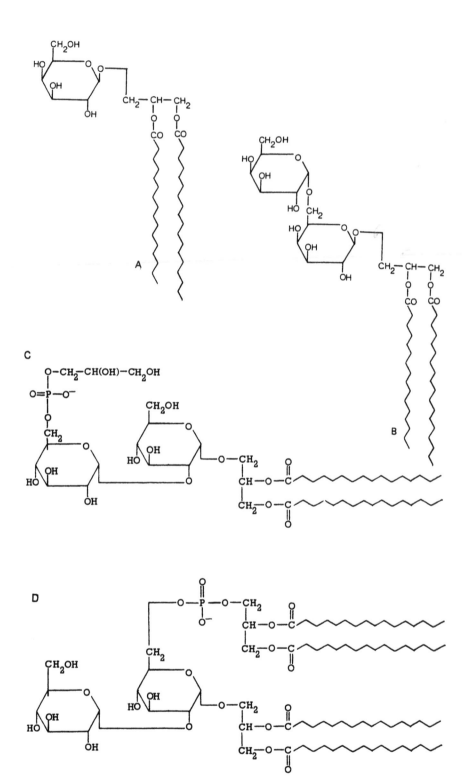

Figure 1.8. Molecular structure of glycolipids based on diacylglycerol: A, monogalactosyldiglyceride; B, digalactosyldiglyceride. Bacterial glycolipids: C, glycerylphosphoryldiglucosyldiglyceride; D, phosphatidyldiglucosyldiglyceride. Reproduced from *Cell Biology Labfax*, Dealtry, G.B. and Rickwood, D. (eds) (1992) with permission from BIOS Scientific Publishers.

Glycosphingolipids. These are only found in animal membranes and consist of ceramide (see *Figure 1.2*) linked (via the free -CH$_2$OH group) to either a straight or branched chain oligosaccharide which often includes N-acetylated sugars and N-acetylneuraminic acid (NANA), sometimes called sialic acid. Most commonly the first sugar is glucose, although in a few cases this is galactose. Some of these structures are shown in *Figure 1.9*. Broadly, there are three groups based on the nature of their oligosaccharide chain:

- Short (less than five neutral residues), unbranched chains.
- Long (five or more neutral residues), mainly branched chains. This group includes blood group active glycosphingolipids; some have as many as 20–50 residues.
- Gangliosides (frequently branched and containing NANA), where normally the number of saccharide residues is 3–7.

The intracellular membranes of eukaryotes are largely devoid of glycolipids which are concentrated in the plasma membrane. In myelin the most common glycosyl ceramide is galactosyl ceramide, which is sometimes sulphated on the C3 position of the sugar. In other plasma membranes, glucosyl ceramide predominates. Many of the gangliosides are also enriched in myelin.

1.3.2 Glycoproteins

Glycoproteins are found predominantly in the plasma membrane where the oligosaccharide chains are exclusively in the extracellular space. Proteins are glycosylated predominantly at asparagine (N-linked) or less commonly at serine or threonine (O-linked) residues and there is often glycosylation at several sites.

N-linked oligosaccharides. The oligosaccharide units consist of core sugars and peripheral sequences which are variable (for examples see *Figure 1.10*). The branched core oligosaccharide (2 GlcNAc and 3 Man) of N-linked structures is derived from a parent core which is both more highly mannosylated and more branched (see Sections 7.2.8 and 7.2.9). The core is linked to the asparagine (Asn) residue through the two GlcNAc residues. *Figure 1.10* shows some of the peripheral sugar sequences which may occur. The main residues are Gal, GlcNAc, NANA and fucose. Some contain repeat sequences of a Gal.GlcNAc disaccharide. More extensive and less commonly, the oligosaccharide chain of N-CAM contains 20–200 NANA residues.

NANA is one of the chief sources of the negative charge on the surfaces of cells. Its presence (or absence) is thus of particular importance in membrane fractionation by continuous flow electrophoresis.

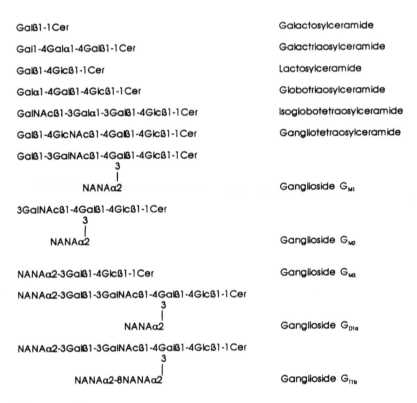

Galß1-1Cer   Galactosylceramide

Gal1-4Galα1-4Galß1-1Cer   Galactriaosylceramide

Galß1-4Glcß1-1Cer   Lactosylceramide

Galα1-4Galß1-4Glcß1-1Cer   Globotriaosylceramide

GalNAcß1-3Galα1-3Galß1-4Glcß1-1Cer   Isoglobotetraosylceramide

Galß1-4GlcNAcß1-4Galß1-4Glcß1-1Cer   Gangliotetraosylceramide

Galß1-3GalNAcß1-4Galß1-4Glcß1-1Cer
                  3
                  |
               NANAα2   Ganglioside G_{M1}

3GalNAcß1-4Galß1-4Glcß1-1Cer
           3
           |
        NANAα2   Ganglioside G_{M2}

NANAα2-3Galß1-4Glcß1-1Cer   Ganglioside G_{M3}

NANAα2-3Galß1-3GalNAcß1-4Galß1-4Glcß1-1Cer
                           3
                           |
                       NANAα2   Ganglioside G_{D1a}

NANAα2-3Galß1-3GalNAcß1-4Galß1-4Glcß1-1Cer
                           3
                           |
           NANAα2-8NANAα2   Ganglioside G_{T1b}

Figure 1.9. Oligosaccharide sequences of some glycosphingolipids.
Reproduced from *Cell Biology Labfax*, Dealtry, G.B. and Rickwood, D. (eds)
(1992) with permission from BIOS Scientific Publishers.

O-linked oligosaccharides. There are five major core sugar sequences
of O-linked oligosaccharides in which the link to a Ser or Thr residue
is via GalNAc (*Figure 1.11*). Each core type is capable of accepting a
huge variety of peripheral sugar sequences which appear to be
controlled by the polypeptide. A number of peripheral sugar sequences
occur, some of which are also shown in the figure. The most common
core sequence is probably type 2. There are, however, some notable
exceptions to the general structure of the oligosaccharide chains;
those in the nuclear membrane for example contain no fucose, GalNAc
or NANA.

1.4.3 Proteoglycans

Proteoglycans consist of long, unbranched carbohydrate chains
usually linked via a serine residue. But, whereas the oligosaccharide
chains of glycoproteins are commonly 7–18 residues long, the

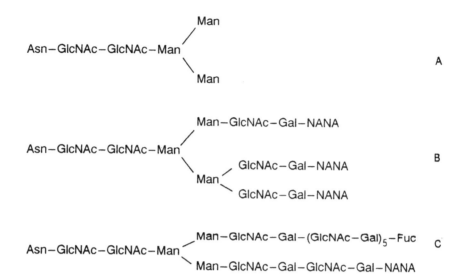

Figure 1.10. N-linked oligosaccharide sequences of mammalian plasma membrane glycoproteins. A, Core oligosaccharide for peripheral sequences; B, core + short peripheral chains; C, core + long peripheral chains. Reproduced from *Cell Biology Labfax*, Dealtry, G.B. and Rickwood, D. (eds) (1992) with permission from BIOS Scientific Publishers.

carbohydrate chains of proteoglycans are frequently 40–100 residues in length and each proteoglycan molecule may contain many of these GAG chains. Some proteoglycans (e.g. most of the heparan sulfate) retain the short transmembrane polypeptide and are thus firmly linked to the membrane, while others lose this link and are bound to surface receptors through ionic interactions.

The GAG chains commonly contain a repeating disaccharide unit, the precise molecular configuration of which may vary from chain to chain. N- or O-sulfated forms of uronic acids (*Figure 1.12*) and amino sugars occur frequently. Some of the common linkage sequences to the core protein are given in *Table 1.7*. Heparan sulfate, dermatan sulfate and chondroitin sulfate occur, which quite widely tend to have high molecular masses; they are commonly between 150 and 450 kD, but may be as high as 900 kD. The structure of heparan sulfate is shown in *Figure 1.13*.

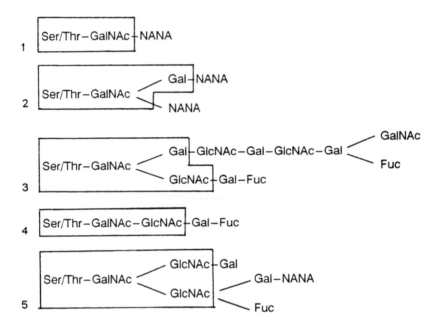

Figure 1.11. O-linked oligosaccharide sequences in glycoproteins. The core oligosaccharide is boxed. Reproduced from *Cell Biology Labfax*, Dealtry, G.B. and Rickwood, D. (eds) (1992) with permission from BIOS Scientific Publishers.

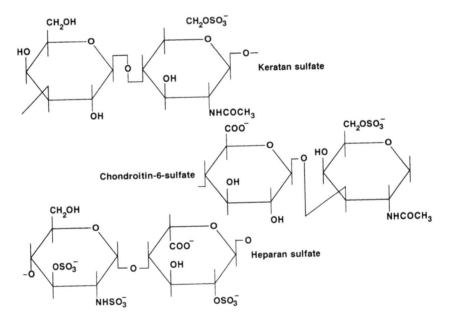

Figure 1.12. Molecular structure of some common proteoglycan disaccharide repeat units.

Table 1.7. Some common oligosaccharide linkage units to the core proteins of proteoglycans

Proteoglycan	Linkage to core protein
Chondroitin sulfate Dermatan sulfate Heparan sulfate Heparin	$-\beta 1-4$GlcUA$\beta 1-3$Gal$\beta 1-3$Gal$\beta 1-4$Xyl$\beta 1-O-$Ser

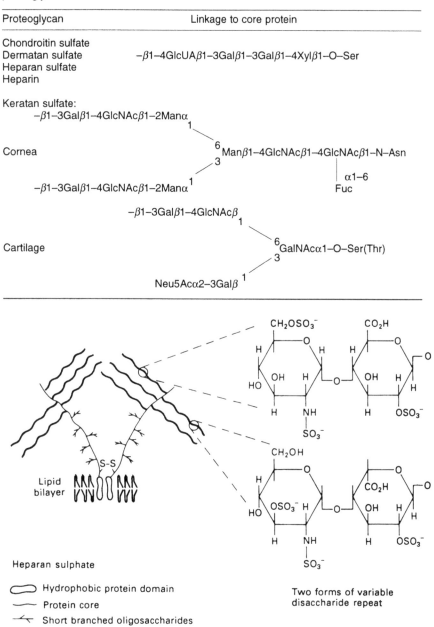

Keratan sulfate:

$-\beta 1-3$Gal$\beta 1-4$GlcNAc$\beta 1-2$Manα

Cornea

$^{6}_{3}$Man$\beta 1-4$GlcNAc$\beta 1-4$GlcNAc$\beta 1-N-$Asn

$\alpha 1-6$
Fuc

$-\beta 1-3$Gal$\beta 1-4$GlcNAc$\beta 1-2$Manα

$-\beta 1-3$Gal$\beta 1-4$GlcNAcβ

Cartilage

$^{6}_{3}$GalNAc$\alpha 1-O-$Ser(Thr)

Neu5Ac$\alpha 2-3$Galβ

CH$_2$OSO$_3^-$ CO$_2$H

OH   H
HO
H   NH
SO$_3^-$

OH   H
H
OSO$_3^-$

CH$_2$OH

CO$_2$H

HO   OSO$_3^-$ H
H   NH
SO$_3^-$

OH   H
H
OSO$_3^-$

Lipid bilayer

S-S

Heparan sulphate

Two forms of variable disaccharide repeat

⊂▢⊃ Hydrophobic protein domain
⁓ Protein core
⤙ Short branched oligosaccharides
〰 Straight glycosaminoglycan chains

Figure 1.13. Molecular structure of heparan sulfate. Reproduced from *Membrane Structure and Function*, Evans, W.H. and Graham, J.M. (1989) with permission from IRL Press at Oxford University Press.

2 The membranes of prokaryotes and eukaryotes

This short chapter is devoted to a overview of the different membrane types and their functions and what consequences there may be for their subsequent fractionation and analysis. It also serves to direct the reader to later sections of the book where functions are described in greater detail and where experimental strategies for their analysis are discussed.

2.1 General features

Membranes are essential components of both prokaryotes and eukaryotes. By definition, the only membrane of prokaryotes is the limiting membrane surrounding the cell; but although in eukaryotes there is a much greater diversity of membranous structures, the function of all membranes can be divided into five principal tasks:

- Membranes form the physical barriers between compartments whose composition can be regulated to provide the proper environments for biological processes to be carried out at high efficiency.
- Membranes control the transport of ions and molecules between compartments to maintain these environments: they are selectively permeable.
- Membranes deliver molecules by a variety of mechanisms into compartments, for their processing.
- Membranes are the receptors for chemical signals and they transduce these signals from one compartment to another.
- Membranes are the site of energy production and the transduction of energy from one form to another.

2.2 Membranes of prokaryotes

Generally in prokaryotes (mainly bacteria) the only membrane is the plasmalemma, sometimes called the cytoplasmic or limiting membrane, which surrounds the organism. The definition does not include the complex layers external to this which may be present in prokaryotes: this outer cell wall provides structural stability. Gram-positive bacteria for example possess a thick (up to 80 nm) layer of complex proteoglycans, while in Gram-negative bacteria this layer is thinner and surrounded by a second membrane which is bimolecular in nature but whose outer leaflet contains lipopolysaccharide molecules, not phospholipids [1].

1. The presence of a cell wall and its nature need to be taken into account when choosing an appropriate homogenization procedure.
2. The cell wall is the site of action of many antibiotics which can therefore be used as part of the homogenization procedure.
3. Sonication is also a common form of homogenization.

The cytoplasmic membrane of the prokaryote has a multifunctional nature; not only is it a selectively permeable membrane responsible for the controlled passage of solute molecules in and out of the cell (like the surface membrane of eukaryotes), but it also contains the enzymes which carry out electron transport and is the site of energy transduction.

- The dominant phospholipid forms in the bacterial cytoplasmic membrane are phosphatidylglycerol (PG), diphosphatidylglycerol (cardiolipin) and phosphatidylethanolamine (PE). Derivatives of PE and PG are also encountered: the CH_2OH group of the glycerol may be linked through an acyl group to an amino acid (amino-acyl ester of PG) and the amine group of PE may be methylated. Mannosyl derivatives of phosphatidylinositol (PI) may also be present (see *Table 2.1* for a typical phospholipid profile). They have usually small amounts of glycolipids, which are highly organism-specific, and there is insufficient space to consider them in detail here. No sterol is present in bacterial membranes. Other distinctive components are hopanes and polyterpenoids (see [2] for more information).
- The membrane of *Escherichia coli* contains more than 200 proteins, 60 of which are probably involved with transport.

Table 2.1. Phospholipids of bacterial membranes[a]

Organism	Phospholipid (% of total phospholipid)[b]					
	PC	PE	NMePE[c]	PG	OAAPG[d]	DPG
Gram-positive						
Micrococcus	<1	<1	<1	60–70	16–18	4–20
Bacillus	<1	20–45	<1	25–45	5–15	10–50
Gram-negative						
Azotobacter	1–2	60–70	3–8	25–30	<1	2–5
Rhodopseudomonas	10–15	45–50	<1	40–45	<1	<1

[a] The values for phospholipid composition vary very widely: average values would be of little use; they depend not only on the type of organism but also on the growth conditions. Generally, PC is absent from Gram-positive bacteria. The outer membranes of Gram-negative bacteria generally contain PE, PG and DPG.
[b] <1 means below the lower limits of detection.
[c] NMePE = N-methylated PE.
[d] OAAPG = O-aminoacyl ester derivative of PG.

Notable highly hydrophobic proteins are the multicomponent BF_1 and BF_0 ATPase complexes.

2.2 Membranes of eukaryotes

Among the eukaryotes, plants, fungi and algae generally have a complex wall composed of celluloses, pectins etc. external to the cell surface membrane which provides an important protection and support for the cells. Although in animal tissue there is a supportive framework of connective tissue, the cells themselves do not have an elaborate wall, indeed the cells of the body fluids (blood, semen etc.) have none whatsoever.

The extent and nature of the extracellular matrix in animal and plant cells determines the mode of homogenization, the severity of which can affect the integrity of the organelles that are released during the procedure.

Compared to prokaryotes, eukaryotes possess a much greater diversity of membrane structures. In addition to the surface membrane (often called plasma membrane in animal cells or protoplast in plant cells) there is a huge range of cytoplasmic structures. Respiratory and photosynthetic mechanisms and energy transduction are delegated to mitochondria and (in plants) chloroplasts. Unlike in prokaryotes, DNA (other than mitochondrial) is restricted to the nucleus, except during meiosis and mitosis.

Lysosomes and peroxisomes are involved with the breakdown of macromolecules and lipid metabolism, respectively, while there is a series of membrane compartments associated with the synthesis and secretion of proteins: the membranes of the rough and smooth endoplasmic reticulum, the Golgi membranes and vast numbers of membrane vesicles which shuttle back and forth between these compartments. Other vesicles are concerned with the uptake of molecules or molecular complexes at the surface for transport to other membrane compartments. A typical animal cell is shown semi-diagrammatically in *Figure 2.1*.

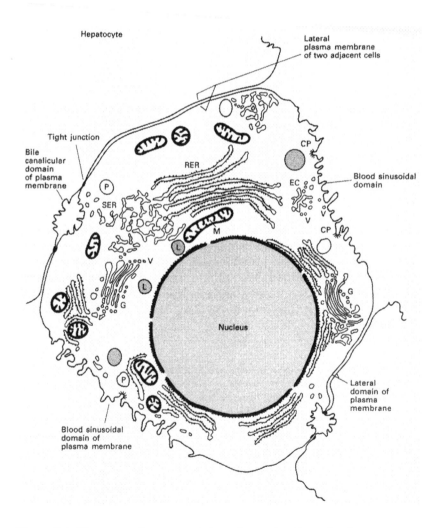

Figure 2.1. Diagrammatic representation of a section through a liver parenchymal cell. Reproduced from Evans, W.H. and Graham, J.M. (1989) with permission from IRL Press at Oxford University Press [18].

2.3 The surface (plasma) membrane

The general composition of the plasma membrane from mammalian cells exhibits the following features:

- The principle phospholipids in plasma membranes are phosphatidylcholine, sphingomyelin and phosphatidylethanolamine, with smaller amounts of phosphatidylserine, phosphatidylinositol and phosphatidylglycerol. High levels of cholesterol are present (0.8–1.0 M ratio to phospholipid). About 10% of total lipid is glycolipid (gangliosides and cerebrosides), although in myelin this may be as high as 28%. See *Table 2.2* for a typical lipid profile of the plasma membrane from both animal and plant cells (for more information see [3] and [4]).
- Seventy or more proteins from the rat liver plasma membrane can be detected by two-dimensional electrophoresis. They include transport proteins for ions (e.g. Na^+/K^+-ATPase); sugars (e.g. the insulin-sensitive glucose transporter) and amino acids; receptors and signal transducers (e.g. low-density lipoprotein receptor, epidermal growth factor receptor, insulin receptor, β_2-adrenergic receptor, G-proteins); and enzymes associated with a variety of functions (e.g. adenylate cyclase).
- Some of the enzymes characteristic of the plasma membranes (and other membranes) from animal tissues (see [5] and [6]) are given in *Table 2.3*.
- The plasma membrane is also enriched in the carbohydrate residues of glycoproteins and glycolipids; NANA (sialic acid) in particular is significantly concentrated at the surface and exposed at the extracellular surface.

Table 2.2. Lipid profile of the plasma membrane

Phospholipid	% of total phospholipid	
	Mammalian cell	Plant cell
Phosphatidylcholine (PC)	35–45	30–35
Phosphatidylethanolamine (PE)	17–22	40–50
Phosphatidylinositol (PI)	6–8	15–20
Phosphatidylserine (PS)	5–10	<1
Phosphatidylglycerol (PG)	2–5	<1
Diphosphatidylglycerol (DPG)	<1	3–5
Sphingomyelin (SM)	15–18	0
Lyso-PC	2–4	na*
Lyso-PE	<1	na*

* Not available.

Table 2.3. Enzymes of mammalian cell membranes

Membrane	Enzyme	Comments
Plasma membrane	5'-Nucleotidase	AMP phosphorylase
	Na⁺/K⁺-ATPase	Inhibited by ouabain
	Alkaline phosphatase	Nonspecific phosphatase
Mitochondria and mitochondrial inner membrane	Succinate dehydrogenase	Most commonly used enzymes of the TCA cycle
	Malate dehydrogenase	
Mitochondrial outer membrane	Monoamine oxidase	Nonspecific
	Rotenone-insensitive NADH–cytochrome c reductase	
Lysosomes	Acid phosphatase	Nonspecific phosphatase
	β-Galactosidase	Two of many glycoprotein degradative enzymes
	β-N-acetylglucos-aminidase	
Peroxisomes	Catalase	Oxidizes hydrogen peroxide
Golgi	Galactosyl transferase	Present in *trans*-Golgi
Endoplasmic reticulum	Glucose-6-phosphatase	
	NADPH-cytochrome c reductase	NADH sometimes used as electron donor
Nuclear membrane	Nucleoside triphosphatase	

1. Enzymes are the most frequently used markers for plasma membrane.
2. High levels of sphingomyelin, cholesterol or NANA can be used as plasma membrane markers. Their measurement is not complicated by the possible inactivation of a functional enzyme activity which may occur during the isolation procedure. The corollary of this is that enzyme markers of course can provide some indication of functional competence.
3. High levels of cholesterol in the plasma membrane can provide a means of density perturbation (see Section 3.2.3) by complexing with digitonin.

Retention of functional activity of plasma membranes (and indeed of all membranes) may be compromised by the release of proteases from lysosomes during the isolation procedure. For this reason a protease inhibitor or a cocktail of protease inhibitors is often included in the homogenization and isolation media (see Section 3.1.5).

The plasma membrane is the site of regulation of all communication between the external environment and the cytoplasm. Ions and small molecules are transported into cells by protein components of the membrane. These include ion channels which are gated by ligands, ions or voltage changes and allow ions and small molecules to move across the membrane down a concentration gradient, and transporters which carry these species across the plasma membrane against a concentration gradient in an energy-dependent manner. Chemical signals including paracrine, endocrine and neurotransmitters interact with receptors, which are also protein components of the plasma membrane. These transduce (change the nature of) the signal-message into the cell. Many of the plasma membrane receptors (which are integral proteins) interact with G-proteins, which in turn produce a second messenger in the cytoplasm by activation/inactivation of an enzyme. Examples of such secondary messengers are cyclic AMP (produced by adenylate cyclase) and inositol trisphosphate and diacylglycerol, produced by a phosphatidylinositol-specific phospholipase C. These second messengers activate kinases which, in turn, affect cell function by phosphorylation of cytosolic proteins. Other receptors are themselves kinases and act directly to phosphorylate cytosolic proteins, when 'turned on' by extracellular signaling molecules.

The study of any of these co-ordinated functions in an isolated membrane fraction is particularly prone to being disturbed by the rather severe means that are often used to homogenize and fractionate cells and tissues. In addition the spatial organization of the intracellular compartment which may be important to many of these complex processes is completely lost. For this reason functional studies on intracellular processes are often carried out in permeabilized cells in which molecular events can be studied with the minimum derangement to the intracellular environment.

2.3.1 Plasma membrane domains

Junctional complexes. There are a variety of junctional complexes which punctuate the plasma membrane of cells from organized tissues (see *Figure 2.2*). **Tight junctions** aid the secretory and absorptive processes: they form belts or gaskets around and between cells, effectively separating the plasma membrane into discrete regions called domains (see [7] and [8]). In this way the surface membrane of the cells of tissues such as intestine and kidney is divided into morphologically and functionally distinct domains (the apical and basolateral domains). In the intestinal epithelial cell (enterocyte) the apical domain is the brush border, across which nutrients pass from the gut lumen. The basolateral domain is adjacent to the blood supply and this membrane functions primarily in the reverse direction, i.e. transporting nutrients from the enterocyte into the blood. Such cells are said to exhibit morphological and functional polarity.

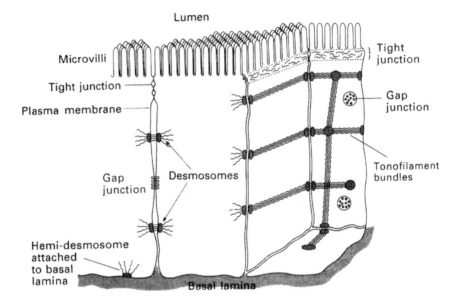

Lumen

Microvilli

Tight junction

Plasma membrane

Tight junction

Gap junction

Gap junction

Desmosomes

Tonofilament bundles

Hemi-desmosome attached to basal lamina

Basal lamina

Figure 2.2. Intercellular junctions in columnar epithelial cells. Reproduced from Evans, W.H. and Graham, J.M. (1989) with permission from IRL Press at Oxford University Press [18].

In the liver, the domain organization of the hepatocyte is further divided: the apical domain is the bile canalicular membrane, while the basolateral domain is divided into the contiguous domain (that part of the basolateral domain adjacent to a neighboring cell) and the blood sinuosidal domain (*Figure 2.1*). Because there is functional specialization at the domain level there is also compositional specialization, notably in the distribution of many enzymes (see *Table 2.4*).

Desmosomes which attach cells together (for example adjacent parenchymal cells in the liver) are attached to tonofilaments which aid in maintaining cell shape and polarity.

Table 2.4. Enzymes of hepatocyte domains

Enzyme	Membrane domain		
	Sinusoidal	Contiguous	Bile canalicular
5'-Nucleotidase	Low	Medium	High
Leucine aminopeptidase	Low	Medium	High
Alkaline phosphodiesterase	Low	Medium	High
Na$^+$/K$^+$-ATPase	Low	High	Low
Glucagon-stimulated adenylate cyclase	High	Low	Low

1. Tight junctions can influence the manner in which the plasma membrane fragments during homogenization, providing 'stress' points at which breakage is likely to occur.
2. Desmosomal structures attached to the plasma membranes between adjacent cells stabilize these membranes against fragmentation.
3. Tight junctions and desmosomes thus influence the surface products of homogenization and any subsequent fractionation scheme to isolate them.

Clathrin. Clathrin and associated proteins (adaptors) may be associated with the cytoplasmic faces of selected areas of the plasma membrane and some membranes of the endocytic and secretory compartments. On the plasma membrane it forms the 'coated' pit at which endocytosis of certain ligands occurs and subsequently surrounds the coated vesicle which is formed (see Chapter 8).

- There is evidence that those areas of the sinusoidal domain engaged in receptor-mediated endocytosis at coated pits are specifically enriched in proteins such as the low density lipoprotein receptor but impoverished in others which are characteristic of that membrane in general such as the adenylate cyclase [9,10].

The presence of clathrin has a significant effect on the density of membrane vesicles with which it is associated and thus influences their fractionation (see Chapter 3).

The plasma membranes of cells which exist primarily within a fluid environment and which do not show any obvious morphological or functional polarity (e.g. blood cells, but not sperm cells) do not possess any obvious domain structure; though surface specialization may still exist in the plane of the membrane. Monolayer culture cells, by definition, show functional polarity, an adherent aspect and a nonadherent surface. Cultured cells such as the MDCK line (canine kidney cells) and Caco-2 cells (human adenocarcinoma line) retain considerable functional polarity [11, 12].

2.3.2 The cytoskeleton and glycocalyx

The **cytoskeleton** forms an important and complex framework in the cytoplasm of all cells. The nature of this framework, however, is extremely variable: it tends to be rather diffuse in most suspension culture cells, while in muscle cells it is highly organized and an integral part of the contractile process. In some cells, for example the human erythrocyte, it forms a mesh just beneath the plasma membrane where, through interactions with integral proteins in the membrane, it may influence cell shape. In cells of some other tissues, e.g. intestinal epithelial cells, this subplasma membrane mesh is considerably elaborated as the terminal web beneath the brush border domain. The cytoskeleton is composed of actin filaments and a number of associated proteins which link actin filaments (villin, fimbrin and myosin) and also form bridges between the cytoskeleton and transmembrane proteins. An example from the human erythrocyte is given in *Figure 2.3*.

The presence or absence and the extent of a cytoskeleton have some important consequences for the homogenization of tissues and cells. The extensive cytoskeleton beneath the enterocyte brush border favors the retention of the entire membrane in an unfragmented state during homogenization. On the other hand, the rather diffuse network in suspension culture cells makes their homogenates difficult to fractionate.

The **glycocalyx** comprises principally a carbohydrate-containing region (as much as 50 nm thick) peripheral to the plasma membrane. Some of the carbohydrate is linked to a protein core within the lipid bilayer (i.e. a transmembrane protein): these molecules may be

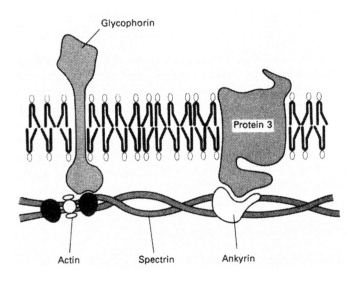

Figure 2.3. Interaction between the cytoskeleton and integral proteins in the human erythrocyte membrane. Reproduced from Evans, W.H. and Graham, J.M. (1989) with permission from IRL Press at Oxford University Press [18].

glycoproteins (e.g. fibronectin) or a proteoglycan such as heparan sulphate (see Section 1.4.3 and [13]), along with more recently identified molecules such as fibronectin, vitronectin and laminin [14]. In plant and fungal cells this extracellular matrix is organized into a discrete cell wall composed of complex polysaccharides which gives support to the internal structures.

- In mammalian cells the molecules of the glycocalyx play an important role in cell–cell adhesion, cell–cell communication and cell proliferation and may be involved in some aspects of signaling.

2.4 Mitochondria and chloroplasts

In animal cells energy transduction occurs exclusively in the mitochondria, while in plants chloroplasts also fulfill this function. Mitochondria are 0.4–2.5 µm in diameter (most are around 1 µm); chloroplasts are larger, approx. 5 µm in diameter. Both organelles have a double limiting membrane, but while the energy transducing reactions of the mitochondrion (the generation of ATP by oxidative phosphorylation) are associated with the infoldings (cristae) of the inner membrane, photosynthesis takes place on the thylakoid membranes within the central stroma and which are separate from the inner membrane. In the chloroplast the inner and outer membrane are collectively termed the envelope. Thylakoid membranes, which may be arranged in stacks or they may form extended lamellae, are responsible for the light harvesting and energy transduction processes and contain the photosynthetic pigments.

The outer membrane of the mitochondrion has some functions (enzymes) and components which resemble the smooth endoplasmic reticulum (SER); they both for example contain low concentrations of cholesterol which is absent from the inner membrane. The inner membrane contains all the components of the electron transport system responsible for the oxidation of tricarboxylic acid cycle intermediates, present in the mitochondrial matrix. It is the energy of the proton gradient generated across this membrane that is the driving force for ATP synthesis. See *Table 2.3* for characteristic enzymes.

Mitochondrial membranes contain complex transporter systems for moving products of the anaerobic breakdown of carbohydrates, lipids and proteins from the cytoplasm into the matrix. A number of these are cyclical mechanisms which exchange one molecule for another, e.g. the glutamate–malate shuttle and the ADP–ATP exchange system. These membranes must also allow the transport and proper insertion of newly synthesized proteins: the process which involves other proteins in the membranes which allow docking, translocation and cleavage of the incoming macromolecules. Although a few inner-membrane proteins (about 10) are manufactured by mitochondrial DNA, the vast majority are imported from the cell synthetic machinery (RER, SER and Golgi apparatus). Investigations into the mechanisms of these processes are a very active area of research (see [15] and [16]).

Plant mitochondria share a lot of the characteristics of their animal counterparts, i.e. they are responsible for the production of ATP, but there are a number of metabolic pathways which are not only particular to plants but specific to they type of tissue and the age of the plant [17]. Consequently some plant mitochondria display properties which are rarely encountered in animal mitochondria, for example CN^--resistance.

- Phospholipids of the outer mitochondrial membrane and the endoplasmic reticulum are broadly similar; phosphatidylcholine (PC) is the major one with smaller amounts of phosphatidylethanolamine (PE) and phosphatidylinositol (PI). The inner mitochondrial membrane is wholly distinctive in being the only eukaryotic membrane with large amounts of diphosphatidylglycerol (cardiolipin), although PC and PE remain the major components. Only the outer membrane has small amounts of cholesterol. The chloroplast membranes comprise mainly PC and phosphatidylglycerol (PG) with small amounts of PI, PE and DPG.

Table 2.5. Phospholipids of mitochondria, chloroplasts and endoplasmic reticulum

Membrane type	Phospholipid as % of total phospholipid						
	PC	PE	PI	PS	PG	DPG	SM
Mitochondrial inner (mammal)	43–48	23–28	5–12	1–2	2–3	16–20	1–2
Mitochondrial inner (plant)	25–35	30–35	3–8	5–25	2–5	15–20	0
Mitochondrial outer (mammal)	50–55	23–25	12–15	2–3	2–3	2–3	3–5
Mitochondrial outer (plant)	40–55	23–28	5–20	8–12	8–12	3–12	0
ER (mammal)	50–60	17–22	8–10	5–10	<1	<1	3–5
ER (plant)	35–60	15–30	10–20	1–2	2–4	2–10	0
Chloroplast envelope	56–75	<1	3–5	<1	20–35	<1	0
Chloroplast lamellae	25–35	<1	<1	<1	60–75	<1	0

The glycolipids of plants are also heavily concentrated in the chloroplast membranes (see *Table 2.5* for details).

Inner membranes of mitochondria are responsible for electron transport and oxidative phosphorylation (*Figure 2.4*). The system is composed of a number of protein complexes, passage of electrons through which is coupled to the movement of protons into the intermembranous space; and two mobile electron carriers, ubiquinone and cytochrome c which shuttle between the complexes. Ubiquinone resides within the lipid bilayer, while cytochrome c is a soluble peripheral protein. Complex 1 (NADH–ubiquinone reductase), Complex III (ubiquinone–cytochrome c reductase) and Complex IV (cytochrome oxidase) are the three sites of proton gradient formation associated with NAD-linked oxidations of TCA cycle intermediates. Complex II (succinate dehydrogenase) is the entry point for the FAD-linked oxidation of succinate. Complex V (ATP synthase) couples ATP synthesis to the transport of protons back into the matrix. It comprises a transmembrane segment as the proton conduit, a stalk (containing the oligomycin binding site) and a globular segment which is the ATP synthetic site (see [18]).

The thylakoid membranes of the chloroplast are the site of a series of complex reactions which are associated with the generation of electron and proton gradients which result in the synthesis of ATP and NADPH. Light-harvesting Complex II delivers energy from light to Photosystem II which splits water and creates the electron and

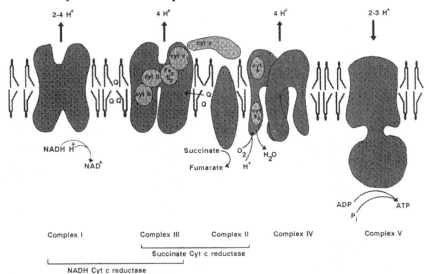

Figure 2.4. Arrangement of proteins in the mitochondrial inner membrane. Complex I is NADH dehydrogenase; Complex II is ubiquinol cytochrome c reductase; complex III is succinate dehydrogenase; Complex IV is cytochrome c oxidase and Complex V is ATP synthase. Q = ubiquinone.

1. The commonly used sources of mitochondria for respiratory studies are beef heart or rat liver. It is critical that their isolation is carried out as quickly as possible. Because of the large numbers of organelles in these tissues, recovery can be sacrificed for purity and simplicity of isolation. Choice of tissue dictates the type of homogenization.
2. Research on the organization and function of mitochondrial membranes often requires the separation of the inner and outer membranes. Intact inner membranes, devoid of the outer membrane are called mitoplasts. The presence of cholesterol exclusively in the outer mitochondrial membrane is the basis for the use of digitonin to remove the outer mitochondrial membrane (see Section 3.2.3).

proton gradients. Electrons are sequentially raised to higher and higher energy levels to allow the synthesis of NADPH by Photosystem I. The whole system involves a series of protein-bound pigments, cytochromes and electron carriers [19].

Chloroplasts are commonly prepared from leaf tissue of peas or spinach. Their tough cell wall, however, requires a quite severe shearing force to release the organelles and it is becoming increasingly common to use plant protoplasts as a material source. Plant protoplasts are the cells of plant tissues from which the wall has been removed by enzymic digestion. For mitochondria it is common to use etiolated plant tissue to avoid interference from chloroplasts during the fractionation [17].

2.5 Peroxisomes (microbodies)

Peroxisomes (the name for these organelles in animals) are smaller than mitochondria (0.4–0.8 µm in diameter) and are surrounded by a single membrane. They are present in a number of animal tissues, primarily liver and kidney where they have been intensively studied. The organelles contain many oxidase functions which act on a broad range of substrates, some of which are tissue-specific, but a general property is that they all produce hydrogen peroxide from oxygen; hydrogen peroxide is also the end-product of the detoxification of oxygen radicals. Approx. 15% of the total peroxisome protein is the enzyme catalase and this enzyme breaks down hydrogen peroxide to water.

In the liver some of the most important reactions of peroxisomes are associated with the β-oxidation of fatty acids. Although the intermediates are identical to those of the β-oxidation which occurs in mitochondria, the enzymes involved are quite different and again one of the end-products when co-factors for the reaction are reoxidized is hydrogen peroxide [20]. Peroxisomes are also involved in the synthesis of ether lipids (plasmalogens).

In plants these organelles are termed microbodies or glyoxysomes. They are responsible for certain metabolic cycles specific to plants. The glyoxylate cycle (glyoxysomes) for example cleaves isocitrate to succinate and glyoxylate which condenses with acetyl CoA to form malate which reforms isocitrate.

> Rat liver is the standard source for peroxisomes and they are purified from a light mitochondrial fraction by density gradient fractionation (see Section 3.4.4).

- The peroxisome membrane itself is has been relatively little studied, although one or two integral peroxisomal membrane proteins have been isolated [21]: research is focused on the enzymes contained within the organelle. There is, however, interest in the manner in which newly synthesized enzymes are targeted to peroxisomes post-translationally from the cytosol (see Chapter 7).

2.6 Lysosomes

Lysosomes are similar in size to peroxisomes and have a single membrane. Through the presence of a proton-translocating ATPase in this membrane the contents are maintained at a pH below that of the surrounding cytosol, normally in the range 5–6. They contain a variety of hydrolytic enzymes (phosphatases, peptidases, esterases and glycosidases) which break down polypeptides, polysaccharides, polynucleotides and glycolipids to yield small molecules which are reutilized by the cell. They form an important part of the endocytic system (see Section 2.8 and Chapter 8); the means by which macromolecules (ligands) are taken up at the cell surface by internalization into membrane vesicles prior to endosome formation. If the ligands are destined for breakdown, then the endosome, after a number of transformations including interactions with other parts of the endocytic compartment, will fuse with a primary lysosome to form a secondary lysosome where digestion of the ligand occurs [22].

Lysosomes are also responsible for the degradation of organelles within the cytoplasm by autophagy. They are able to engulf, by invagination around, other organelles and then, following fusion between the membranes of the lysosome and those of the organelle, the latter is digested by the hydrolytic enzymes.

1. Although rat liver has been a common source of lysosomes for structural and functional studies, the wide range of cells and tissues used in the studies of endocytosis and membrane trafficking means that these organelles have been purified from yeast and cultured mammalian cells, particularly the polarized ones such as MDCK or Caco 2 cells.
2. Lysosomes are routinely purified from a light mitochondrial fraction by density gradient fractionation (see Section 3.4.4).

2.7 Nuclei

Other than during meiosis and mitosis, the genetic material of eukaryotes is enclosed by a double-layered nuclear membrane. It is not, however, a true double membrane, in the sense that a double membrane surrounds the mitochondrion. In the nuclear membrane, there are large well defined pores which regulate movement of macromolecules into the cytoplasm: at these pores electron microscopy shows that there is continuity between the inner and outer membranes. It is therefore more useful to think of the nuclear membrane as flattened membrane-bound lamellae. Chromatin is attached to the inner membrane, while the outer membrane often has attached ribosomes and is continuous with the RER (see *Figure 2.5*).

Beneath the inner aspect of the membrane is a felt-like protein, comprising the nuclear lamina. This is continuous with the nuclear pore complexes. These are intricate associations of proteins arrayed in a series of 'radiating spokes'. All proteins and ribosomal subunits moving between the nucleus and the cytoplasm must traverse the nuclear pores: there is therefore considerable research interest in the mechanism regulating this process and the detailed structure and function of the nuclear pore proteins and their associated structures (*Figure 2.5*). See [23] for more information.

• The only specific enzymic function associated with the membrane appears to be a nucleoside triphosphatase. Other enzymes present

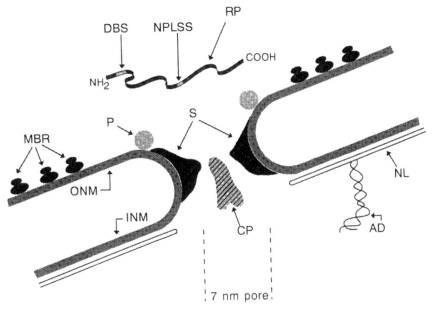

Figure 2.5. Nuclear pore structure. ONM = outer nuclear membrane; INM = inner nuclear membrane; MBR = membrane-bound ribosome; P = particle (one of octet around pore); CP = central plug; AD = attached DNA; NL = nuclear lamina; S = spoke; RP = regulator protein; DBS = DNA binding sequence; NPLSS = nuclear pore localization signal sequence.

are representative of the ER and may be true components or merely reflect the presence of some attached ER membranes.

1. Rat liver is the common source of large scale preparations of nuclei. Nuclei from mammalian cultured cells are often extremely fragile and tend to release their DNA rather easily. Plant nuclei are frequently isolated from wheat-germ to avoid the problem of chloroplast contamination from green plants.
2. The isolation of the nuclear membrane is made difficult by the necessity to remove and solubilize large amounts of DNA (see Section 3.4.1).

2.8 Endocytic compartment

Endocytosis is a multi-step process [22,24] by which a ligand bound to a receptor at the plasma membrane (or plasma membrane domain) is internalized and translocated to lysosomes, to another domain of the plasma membrane (transcytosis), back to the same domain from which it was internalized (diacytosis) or to some other internal compartment. The initial internalization at the plasma membrane may or may not involve clathrin-coated pits. Internalization at a coated pit results in the formation of a coated vesicle, from which a primary endosome is formed; while internalization outside a coated pit forms a primary endosome directly. Primary endosomes deliver the ligand to a sorting station – the tubular early endosome compartment – which may become multi-vesicular with the formation of internal vesicles prior to delivery of the receptor (and/or ligand) to one of the four destinations. Alternatively carrier vesicles may bud off the early endosome compartment which translocates the ligand to late endosomes. Hydrolysis of the ligand may take place in a late endosomal (pre-lysosomal) compartment. The process is summarized and its analysis covered in Chapter 8 (see also *Figure 8.1*).

1. Endocytosis is commonly studied using either a perfused liver system or cultured cells. The former is closer to the physiological state, but only a single time point can be investigated per experiment. Cultured cells are much easier to handle and many time points can be investigated in a single experiment but the system is more 'removed' from the physiological state.
2. Ligands can be labelled radioisotopically, with colloidal gold, with a fluorescent tag or with horseradish peroxidase. Choice of marker will depend on whether the subsequent analysis of the endocytic process is carried out either by centrifugation in density gradients, by light microscopy or by electron microscopy (see Chapter 8).

2.9 The secretory compartment

The secretory compartment is part of the more general membrane synthetic compartment and is involved in the very complex trafficking of membrane material around the cell. It comprises five main compartments: the rough endoplasmic reticulum (RER), smooth endoplasmic reticulum (SER), *cis*-Golgi network (CGN), the Golgi apparatus and the *trans*-Golgi network (TGN). The Golgi apparatus is futher divided on positional and functional grounds as *cis*-, medial- and *trans*-Golgi. The translocation of macromolecules through this sytem involves, in addition to these major membrane compartments, a number of carrier vesicles which shuttle between some of them (see Chapter 8).

The RER is the site of synthesis of membrane proteins and proteins destined for secretion. Newly synthesized proteins are co-translationally and post-translationally modified in their passage through the components of the secretory (synthetic) system. Protein modification includes glycosylation to form glycoproteins and proteoglycans, acylation, isoprenylation, phosphorylation and attachment of the protein to PI. Analysis of these complex processes is dealt with in Chapter 7. Unlike endocytosis, at least some of the major secretory compartments have quite distinctive functional markers. In addition to those listed in *Table 2.3*:

- the SER is characterized generally by β-D-glucosidase which is involved in the early processing of glycoproteins;
- the RER is characterized by its RNA content;
- the ER and in particular the ER of muscle cells where it is called sarcoplasmic reticulum is involved with the regulation of intracellular Ca^{2+} levels, and Ca^{2+} ions are pumped out of the cytoplasm into the cisternae of the reticulum by the Ca^{2+}-ATPase in this membrane.

1. Radiolabeled precursors of proteins, glycoproteins and proteoglycans can be used to follow specific macromolecules through the secretory system by classical pulse-chase analysis. This can be carried out using a perfused liver system or rather more conveniently a cultured cell system.
2. The identification of specific compartments in the secretory process can be achieved by enzyme analysis if applicable (see Section 3.3.3) or the use of antibodies to specific proteins.

Having reviewed very briefly, in Chapters 1 and 2, the types of membranes that exist in prokaryotes and eukaryotes and some of their functions, properties and composition; we are now ready in the next two chapters to discuss the means of fractionating the different membrane types; how we can use some of their functional characteristics to monitor this fractionation by use of the appropriate membrane markers and the means by which these markers and other membrane molecules are analysed.

References

1. **Rogers, H.J., Perkins, H.R. and Ward, J.B.** (1980) *Microbial Cell Walls and Membranes.* Chapman and Hall, London.

2. **Poole, R.K.** (1993) in *Methods in Molecular Biology,* Vol. 19 (J.M. Graham and J.A. Higgins, eds), pp. 109–122. Humana Press, Totowa, NJ.
3. **Hawthorne, J.N. and Ansell, G.B.** (1982) *Phospholipids.* Elsevier North Holland, Amsterdam.
4. **Harwood, J.L. and Walton, T.J.** (1988) *Plant Membranes – Structure, Assembly and Function.* Biochemical Society, London.
5. **Graham, J.M.** (1992) in *Cell Biology Labfax* (G.B. Dealtry and D. Rickwood, eds), pp. 77–101. BIOS Scientific Publishers, Oxford.
6. **Graham, J.M.** (1993) in *Methods in Molecular Biology,* Vol. 19 (J.M. Graham and J.A. Higgins, eds), pp. 1–18, Humana Press, Totowa, NJ.
7. *Junctional Complexes of Epithelial Cells (1987) Ciba Foundation Symposium 125,* John Wiley and Sons, Chichester.
8. **Gumbiner, B.** (1987) *Am. J. Physiol.* **253:** C749–C758.
9. **Pearse, B.M.F. and Crowther, R.A.** (1987) *Annu. Rev. Biophys. Biophys. Chem.* **16:** 49–68.
10. **Moore, M.S., Mahaffey, D.J., Bordsky, F.M. and Anderson, R.G.W.** (1987) *Science* **236:** 558–563.
11. **Bartles, J.R. and Hubbard, A.L.** (1988) *Trends Biochem. Sci.* **13:** 181–184.
12. **Matter, K. and Mellman, I.** (1994) *Curr. Opin. Cell Biol.* **6:** 545–554.
13. **Gallagher, J.T., Lyon, M. and Steward, W.P.** (1986) *Biochem. J.* **236:** 313–325.
14. **Chambers, J.A.A.** (1993) in *Biochemistry Labfax* (J.A.A. Chambers and D. Rickwood, eds), pp. 167–191. BIOS Scientific Publishers, Oxford.
15. **Lill, R., Nargang, F.E. and Neupert, W.** (1996) *Curr. Opin. Cell Biol.* **8:** 505–512.
16. **Haucke, V. and Schatz, G.** (1997) *Trends Cell Biol.* **7:** 103–106.
17. **Moore, A.L. and Whitehouse, D.G.** (1997) in *Subcellular Fractionation – A Practical Approach* (J.M. Graham and D. Rickwood, eds), pp. 243–270. IRL Press at Oxford University Press, Oxford.
18. **Evans, W.H. and Graham, J.M.** (1989) in *Membrane Structure and Function,* pp. 70–82. IRL Press at Oxford University Press, Oxford.
19. **Rochaix, J.D. and Erickson, J.** (1988) *Trends Biochem. Sci.* **13:** 56–59.
20. **Van den Boscg, H., Schutgens, R.B.H., Wanders, R.J.A. and Tager, J.** (1992) *Annu. Rev. Biochem.* **61:** 157–197.
21. **Völkl, A., Baumgart, E. and Fahimi, H.D.** (1997) in *Subcellular Fractionation – A Practical Approach* (J.M. Graham and D. Rickwood, eds), pp. 143–167. IRL Press at Oxford University Press, Oxford.
22. **Gjøen, T., Berg, T.O. and Berg, T.** (1997) in *Subcellular Fractionation – A Practical Approach* (J.M. Graham and D. Rickwood, eds), pp. 169–203. IRL Press at Oxford University Press, Oxford.
23. **Dingwall, C. and Laskey, R.A.** (1986) *Annu. Rev. Cell Biol.* **2:** 367–390.
24. **Murphy, R.F.** (1991) *Trends Cell Biol.* **1:** 77–87.

3 Preparation of subcellular membranes

There are two principal steps to the preparation of subcellular membranes: homogenization of the tissue or cells (Section 3.1), followed by fractionation (or separation) of the organelles and membranes (Section 3.2). This fractionation process is such a vast and complex area that it is not possible to provide more than a few basic themes. An essential part of the fractionation process is the identification of the organelles and membranes so as to assess the efficacy of the procedure and this is covered briefly in Section 3.3, while Section 3.4 contains a few illustrative examples of common fractionations.

3.1 Homogenization

It should always be borne in mind that as soon as cells are broken, they have undergone a serious assault on their integrity; the organelles are subjected to forces and exposed to media that they never encounter *in vivo*. Moreover the cells which (at least in the case of mammalian tissues) have been accustomed to operating at approx. 37°C, are normally rapidly cooled to 0–4°C for homogenization and the ensuing fractionation process. Such a significant fall in temperature will certainly lead to changes in membrane lipid fluidity and may cause a change in protein function. The use of low temperatures is, however, unavoidable. Although the aim of the homogenization process is to rupture all the surface membranes of all the cells and so release all of the intracellular organelles in an intact state, this ideal situation is never realized. Some of the organelles will themselves be ruptured, so releasing enzymes which *in vivo* are sequestered within an organelle. As some of these enzymes are degradative, such as proteases and lipases, low temperatures are essential to limit enzyme activity and it is often recommended that inhibitors of these enzymes (particularly proteases) are included in the homogenization (and all subsequent) media.

In spite of these problems, homogenization and the ensuing membrane fractionation are often the only means of studying the function and structure of a specific subcellular particle. Moreover, the recovery in isolated particles of functional activities (biochemical, immunological and physiological) known to be associated with those organelles *in vivo* attests to the validity of the procedure. Some of the most commonly used homogenization techniques as applied to eukaryotic cells and tissues are described in Sections 3.1.1–3.1.5 and subsequent sections will describe how these methods are applied to various types of tissue and cell. See [1] for a detailed discussion about homogenization.

3.1.1 Liquid shear

Dounce and Potter–Elvehjem homogenizers. Liquid shear is routinely generated by a liquid flowing through the narrow space between the glass body of the homogenizer and a moving pestle. There are two basic designs of this type of homogenizer. The pestle of the manually operated Dounce (sometimes called 'all-glass') homogenizer (*Figure 3.1*) comprises a glass ball and the liquid shear force depends on the thrust of the pestle and the clearance between it and the container. Clearances are normally 0.03–0.08 mm (tight-fitting) or 0.1–0.3 mm (loose-fitting). The former are commonly called 'Wheaton Type A' and

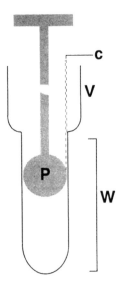

Figure 3.1. Dounce homogenizer. P, Glass pestle; V, glass containing vessel; W, working volume; c, clearance between pestle and vessel of 0.03–0.3 mm.

the latter 'Wheaton Type B'. The pestle of a Potter–Elvehjem (sometimes called 'Teflon and glass') homogenizer (*Figure 3.2*) is normally a solid cylinder, hemispherical at its end, of Teflon (occasionally this is ball-shaped) and the liquid shear force is additionally controlled by the speed of rotation of the pestle (normally 500–1000 r.p.m.), which is attached to some form of overhead motor (high torque and thyristor controlled). The clearance between the pestle and the glass vessel is normally 0.08–0.3 mm.

In both types of homogenizer the cumulated shear force experienced by the biological material depends also on the number of passes of the pestle. For the Potter–Elvehjem homogenizer this is normally up to eight (up-and-down) strokes and for the Dounce, up to 20 strokes.

Other liquid shear devices. An alternative to the Dounce homogenizer is the 'Cell Cracker' [2] in which the cell suspension is forced repeatedly from fixed syringes through a 0.03 mm gap between a stainless steel ball and a precision bore in a stainless steel block. Two other simple alternatives involve either repeated drawing into and expulsion from syringes through narrow bore needles or through a stainless steel screen (pore size 110 µm).

In all of these devices, the shear force has to be applied either continuously or for a number of cycles. They suffer from two particular disadvantages:

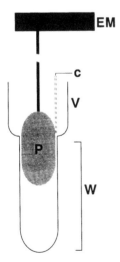

Figure 3.2. Potter–Elvehjem homogenizer. EM, Electric motor; for other abbreviations see *Figure 3.1*.

- The shear force is generated by the operator and is impossible to quantify and hence not very reproducible.
- Some of the cells will be broken and their contents released early in the process, others later on: the released organelles themselves will thus be exposed to varying comminution forces. This is at least partially responsible for the diversity of sizes of some membrane particles and rupture of some organelles.

Some of the problems asociated with the manual devices are overcome if the shear force is applied mechanically. In the 'Stansted' cell disruptor (*Figure 3.3*) the cell suspension is forced under pressure by the piston of a compressed air pump through an orifice. The shearing forces are controlled by the size of the orifice and the air pressure (normally 100–150 MPa). Usually a single pass of the cells is sufficient to achieve good breakage, sometimes two are necessary.

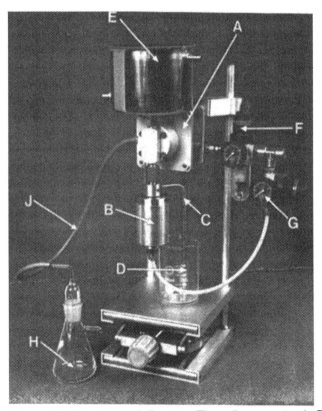

Figure 3.3. Stansted cell disruptor. The cell suspension in E is forced by the pump (A) through the valve unit (B) using compressed air and controlled by the flow and pressure regulators (F and G). The homogenate is collected in beaker D via the delivery tube (C). An air outlet (J) allows exhaust air to pass to a sterilizing solution in flask H. Reproduced from Evans, W.H. (1992) with permission from IRL Press at Oxford University Press [3].

3.1.2 Gaseous shear

In nitrogen cavitation (*Figure 3.4*) the cell suspension is placed in a stainless steel pressure vessel. It is first equilibrated at 4°C with oxygen-free nitrogen at about 5500 kPa for 10–30 min and using the gas pressure it is expelled via a delivery tube through a needle valve. Here it is exposed to atmospheric pressure and cell homogenization occurs due to the rapid expansion of nitrogen gas dissolved within the cytosol and to the rapid formation of bubbles in the medium.

The advantage of both nitrogen cavitation and the Stansted machine is that cell disruption occurs instantaneously and the released organelles are not exposed to a continuing and variable shearing force. Moreover in nitrogen cavitation, the expansion of the N_2 causes cooling, while all other forms of shear tend to generate heat.

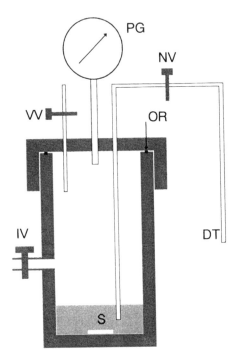

Figure 3.4. Nitrogen cavitation. The sample is stirred with a magnetic stirrer (S) while being equilibrated with oxygen-free nitrogen which enters at the inlet valve (IV); the pressure is monitored by the gage (PG). Homogenization occurs when the sample is extruded at the needle valve (NV) and the homogenate is collected in a beaker via the delivery tube (DT). VV, Vent valve; OR, O-ring.

3.1.3 Mechanical shear

Mechanical shear is achieved in some sort of rotating blades homogenizer (not unlike the domestic blender or liquidizer). The Waring Blender and Ultra-Turrax were two of the earliest commercially available machines and generally deal with quite large volumes of material (>100 ml). The Waring Blender resembles very closely the domestic liquidizer, while the Ultra-Turrax comprises a cylindrical metal tube containing a set of rotating teeth. A modern sophisticated version of the Ultra-Turrax is the Polytron homogenizer which is shown in *Figure 3.5*; the tip of the metal tube can be attached to a range of work-heads, so that it can be adapted to volumes as low as 0.2 ml and as high as 5 l.

With Waring Blender-type homogenizers, the vortex created in the whole sample can cause serious frothing and consequent protein denaturation. In the Polytron-type the sample is continuously drawn into and expelled from the work-head, making the disruptive process more effective, more reproducible and less prone to frothing.

Mechnical shear is sometimes used to produce a 'coarse' homogenate as a prelude to a finer liquid shear homogenization in a Potter–Elvehjem or Dounce homogenizer.

3.1.4 Other methods

Sonication is generally too drastic to be applied to eukaryotic cells such that only the surface membrane is ruptured, without breaking the internal organelles. Local heating at the probe tip can also be a problem. If this technique is used, the force is applied only for short (often 10 sec) 'bursts' with 'rest' intervals of 30 sec. Sonication is only widely used for prokaryotes. Its only significant application to eukaryotes is for the preparation of vesicles from large sheets of plasma membrane.

Osmotic lysis occurs when a cell suspension is exposed to a hypo-osmotic medium so that water rapidly enters the cell and causes bursting. The technique is simple and highly reproducible, but consequent exposure of the organelles to the same low osmolarity is a problem, although this can be minimized by using as small a volume of medium as possible so that the cytosolic proteins afford some osmotic protection (see Section 3.1.7). Occasionally a low osmolarity solution is used merely to swell the cells to render them more susceptible to liquid shear, especially in the case of isolated cells or suspension culture cells.

Figure 3.5. Polytron homogenizer. The inset shows the rotating teeth within the work-head. Reproduced from Evans, W.H. (1992) with permission from IRL Press at Oxford University Press [3].

3.1.5 Homogenization of soft mammalian tissue

Homogenization media. The medium is normally iso-osmotic (that is it has an osmolarity approximately equal to that of the cytosol) and contains a polyhydric alcohol; commonly this is 0.25 M sucrose (although for brain 0.32 M sucrose is used) together with an organic buffer (normally at 10–20 mM concentration): Tris–HCl buffers are the most common but Hepes or Tricine buffers, which are generally less toxic than Tris, are becoming more widely used. Buffers, however, are not essential as the proteins of the cytosol provide adequate buffering capacity. Some of the commonly used media and their uses are given in *Table 3.1*. The presence of divalent cations (Mg^{2+} and Ca^{2+}) is generally avoided in routine isolation media as they tend to cause aggregation of membranes and Ca^{2+} activates certain phospholipases and proteases. For this reason either EDTA or EGTA (1 mM) are added to homogenization media.

- In the case of nuclei, a chelator, which might induce fragility of these organelles, is omitted and replaced by Mg^{2+}.
- Large sheets of liver contiguous plasma membrane may also be stabilized by Mg^{2+}.
- KCl is also a common supplement for nuclei, mitochondria and in a general purpose medium when it may be necessary to solubilize proteins which may interfere with the subsequent fractionation.

A routine homogenization schedule for mammalian liver for general organelle isolation is given in Box 3.1.

Box 3.1

Aim: Homogenization of rat liver.

Protocol

Step 1. Place the excised tissue in a beaker on ice and process into a fine mince using scissors (pieces 1–2 mm³ in size).

Step 2. Suspend the mince in the homogenization buffer (approx. 40 ml for 10 g of tissue) and transfer half of the tissue suspension to an ice-cold 20–25 ml Potter–Elvehjem homogenizer.

Step 3. Attach the ice-cold pestle to a suitable overhead electric motor and introduce into the homogenizer barrel. Homogenize using five to six up-and-down strokes of the pestle rotating at 700 r.p.m.

Step 4. Repeat the procedure for the other half of the liver mince.

Points to note

It is easier to work with 10–12 g of liver in two batches using a 20–25 ml homogenizer rather than one batch in a larger 40 ml one.

During the first downward stroke of the pestle, some of the material will tend to compact at the bottom of the glass barrel. Never force the pestle through this, instead use the vortex action of the pestle during the up stroke to resuspend this material.

General applicability: Liver, kidney, intestinal mucosa and brain tissue.

- To avoid contamination by erythrocytes of sheets of plasma membrane, isolated from the nuclear pellet, a highly vascular

Table 3.1. Homogenization media for soft mammalian tissues

Tissue	Medium	Uses
Liver	(1) 0.25 M sucrose, 1 mM EDTA, Hepes–NaOH, pH 7.4	General purpose medium
	(2) 0.2 M mannitol, 50 mM sucrose, 1 mM EDTA, 10 mM Hepes–NaOH, pH 7.4	Mitochondria
	(3) As (1) plus 0.1% ethanol	Peroxisomes
	(4) 0.25 M sucrose, 25 mM KCl, 5 mM MgCl$_2$, 20 mM Tris–HCl, pH 7.4	Nuclei
Brain	0.32 M sucrose, 1 mM EGTA, 0.5 mM MgCl$_2$, 20 mM MES–NaOH, pH 6.5	Synaptosomes

tissue such as liver should be perfused with ice-cold homogenization medium to remove as much of the blood as possible prior to homogenization. *Always seek expert assistance from a trained person in your animal unit before carrying out a perfusion.*

- For many applications it is desirable to include a cocktail of protease inhibitors in the medium to minimize degradation due to these enzymes (see *Table 3.2*). For more information see [3] and [4].

- If the 1000g pellet is to be processed, the homogenate should be filtered through nylon mesh (75 μm pore size) to remove unbroken clumps of cells and connective tissue which would contaminate this pellet.

Table 3.2. Commonly used protease inhibitors

Inhibitor	Specificity	Stock solutions	Final conc.
Amastatin	Amino exopeptidases	Ethanol	1–10 mM
Antipain	Cathepsin B, papain, trypsin	DMSO[a]	1–100 mM
Aprotinin	Serine proteases	Water	<2.0 μg ml^{-1}
Benzamidine	Serine proteases	Ethanol	0.5–4.0 mM
Chymostatin	Cathepsin B, chymotrypsin, papain	DMSO[a]	10–100 μM
DFP[b]	Serine proteases	Propan-2-ol	100 μM
EDTA	Metalloproteases	Water	0.1–5.0 mM
Leupeptin	Trypsin-like and cysteine proteases	Water	10–100 μM
Pepstatin A	Cathepsin D and G, pepsin, renin	Methanol	1 μM
PMSF[c]	Serine proteases	Ethanol	0.1–1.0 mM
Phosphoramidon	Collagenase, thermolysin	DMSO[a]	1–10 μM

[a]DMSO = dimethylsulfoxide; [b]DFP = diisopropylphosphofluoridate (extremely toxic);
[c]PMSF = phenylmethylsulfonyl fluoride (short half-life but very useful general inhibitor).

3.1.6 Homogenization of tough tissues

A tissue such as skeletal muscle, normally too tough to be homogenized by liquid shear, can be homogenized using a mechanical shear method. However, because of the possible damage to released organelles during prolonged homogenization, the finely cut tissue is often incubated with a protease (the most popular one is the commercial 'Nagarse') to produce partial disaggregation of the tissue (softening) before homogenization by a short period of mechanical shear and/or by liquid shear.

Homogenization media. Although organic buffers containing polyhydric alcohols similar to those described in Section 3.1.5 may be used for tough tissues, salt buffers are not uncommon, especially if the homogenization of the tissue releases gelatinous material that may cause the aggregation of membranes. Muscle tissue is often homogenized in 100–150 mM KCl and the well-known Chappel–Perry medium [5] which is widely used for the preparation of skeletal muscle mitochondria contains 100 mM KCl, 20 mM MOPS, 5 mM $MgSO_4$, 1 mM ATP, 5 mM EDTA and 0.2% bovine serum albumin (BSA). A typical homogenization is given in Box 3.2.

Box 3.2

Aim: Homogenization of skeletal muscle.

Protocol

Step 1. Wash the tissue in a buffered mannitol/sucrose/EDTA medium and then mince very finely using two scalpels or razor blades. A very fine mince is critical for the following enzyme digestion.

Step 2. Incubate with Nagarse (0.1–0.2 mg ml^{-1}) in 0.1 M sucrose, 10 mM EDTA, 40 mM KCl, 0.5% (w/v) BSA and 100 mM Tris–HCl, pH 7.4, at 0–4°C for about 5 min.

Step 3. Homogenize in a Potter–Elvehjem homogenizer (see Box 3.1).

Step 4. Repeat the incubation and rehomogenize after dilution with an equal volume of buffer (without enzyme).

General applicability: Skeletal muscle from most experimental animals.

3.1.7 Homogenization of cultured mammalian cells

Liquid shear methods. Monolayer cells such as fibroblasts pose relatively few problems for homogenization; generally they can be disrupted in a tight-fitting Dounce homogenizer in an iso-osmotic sucrose medium containing 1 mM EDTA (Box 3.3). The buffer that is used, however, appears to be critical and 10 mM triethanolamine–10 mM acetic acid, pH 7.6, is superior to all other buffers [6,7].

Box 3.3

Aim: Homogenization of monolayer cells.

Protocol

Step I. Remove the growth medium completely by careful washing of the monolayer with phosphate buffered saline (at room temperature); at least three washes are recommended: the presence of serum proteins and divalent cations from the growth medium severely inhibits homogenization.

Step 2. Wash the monolayer at least twice with 0.25 M sucrose, 10 mM triethanolamine, 10 mM acetic acid, pH 7.4, to remove the saline wash.

Step 3. Scrape the monolayer into 2 ml of ice-cold 0.25 M sucrose, 10 mM triethanolamine, 10 mM acetic acid, 1 mM EDTA, pH 7.4. Break up the monolayer into small 'plaques' not into single cells.

Step 4. Process the cells in batches of three using a 6 ml tight-fitting Dounce homogenizer. Normally about 20 strokes of the pestle are required to effect 90–95% cell breakage. Sometimes as many as 30 strokes are required.

Step 5. Monitor the homogenization by phase-contrast microscopy.

General applicability: Near-confluent monolayers of any cultured cell; fibroblastic cells respond the best. If more than 30 strokes of the pestle are needed to achieve 90% breakage, use another method (see Box 3.4).

Suspension culture cells do not usually disrupt using liquid shear in an iso-osmotic medium. For such cells to become susceptible to homogenization, they must be swollen. Generally the strategy is to use as small a volume of medium as possible so that when the cells do break, the cytosolic proteins are at a sufficiently high concentration to protect the released organelles from the hypo-osmotic conditions. Once satisfactory homogenization is achieved, concentrated sucrose is then added to make the final concentration 0.25 M. If using a small volume of hypo-osmotic medium produces insufficient swelling, then the procedure used by Goldberg and Kornfeld [8] and Dunphy and Rothman [9] is useful (Box 3.4).

Other hypo-osmotic solutions that have been used include 1 mM $NaHCO_3$, 0.1 M sucrose, and 10–20 mM of an organic buffer. The presence of EDTA is best avoided as the nuclei, which tend to be rather fragile in hypo-osmotic media, would probably break down entirely. More reproducible results with liquid shear of cells can be obtained using either the Cell Cracker or the Stansted machine (see Section 3.1.1).

Box 3.4

Aim: Homogenization of suspension culture cells [8,9].

Protocol

Step 1. Suspend the cells in a relatively large volume of medium (approx. 2×10^8 cells in 10 ml); 15 mM KCl, 1.5 mM Mg-acetate, 1 mM dithiothreitol and 10 mM Hepes–KOH, pH 7.5, is a good medium.

Step 2. Allow them to swell for 5–10 min at 0°C, then sediment them gently and remove at least 5 ml of the medium.

Step 3. Homogenize the cells in the residual volume of buffer. The volume of residual buffer needs to be about 3.5× that of the cells.

Step 4. Add a concentrated sucrose solution to restore iso-osmotic conditions.

Nitrogen cavitation. Any cell suspension can be homogenized effectively by nitrogen cavitation in a routine iso-osmotic medium (buffered 0.25 M sucrose). It is extremely reproducible and can be used for a wide range of sample volumes (10 ml to 1 l). It is not uncommon to use 2–6×10^7 cells/ml of medium and over this range, the efficiency of homogenization is independent of cell concentrations. One possibly deliterious effect of nitrogen cavitation is that ribosomes tend to be stripped off of the rough endoplasmic reticulum.

- The nuclei of cultured cells seem particularly fragile and homogenization (particularly in hypo-osmotic media) can lead to the release of DNA. DNase I is therefore often added to overcome this problem. In severe cases visible aggregates will form and a low speed centrifugation to pellet the nuclei may cause virtually all the cellular material to sediment.
- Another problem with cultured cells is that any cytoskeleton tends to be rather less well developed and localized compared to that in cells from organized tissues. After homogenization it may become dispersed rather than remaining with the membrane and cause aggregation.

3.1.8 Plant tissue

There are two approaches to the homogenization of plant tissue: a consequence of the tough cell wall material surrounding plant cells. One is to use quite severe mechanical means (a blender or grinding with silica or glass beads), the other is to digest the cell wall with enzymes to release the plant protoplasts (plant cell minus cell wall) which then require only very gentle means of cell rupture. Because chloroplasts in particular are very sensitive to shearing forces protoplast formation has become a popular method. Degradative enzymes and toxic phenolic compounds can also be released from disrupted organelles. The presence of large amounts of starch grains can also lead to damage to chloroplasts during homogenization [10], and for this reason it is common to isolate chloroplasts from leaves at the start of the light period.

Mechanical shear. A hand-held domestic liquidizer is often used to achieve a partial maceration of the leaf tissue, followed by further mechnical shearing with an Ultra-Turrax or Polytron homogenizer. To avoid the problem of organelle damage, incomplete disruption of all the tissue is often accepted, as the homogenate is always extensively filtered through several layers of muslin or cheesecloth to remove larger particles [10].

Homogenization media. It is common to use quite large amounts of starting material (e.g. 50–100 g of leaves) and volumes of homogenization buffer from 150 to 600 ml. A medium containing buffered sucrose, mannitol or sorbitol (at 0.3 M concentration) is perhaps the most common [10]. The precise composition of the medium, however, depends on the organelle of interest and the plant species and will be considered later. One interesting point is that homogenization media are often used as a semi-frozen 'slush' rather than an ice-cold liquid. A complete discussion of these fine points is beyond the scope of this chapter and the reader is referred to a number of excellent reviews [10,11].

Plant protoplasts. The production of plant protoplasts from leaves often requires considerable pre-treatment of the tissue to allow sufficient access of the digestive mixture to as much of the cellular material as possible. Abrasion with carborundum or fine cutting with a razor blade are common techniques. One problem that may be encountered is that during this pre-treatment, air may enter the tissue and prevent complete exposure of the cells to the enzyme mixture, in which case cutting is sometimes carried out under a liquid or vacuum infiltration used to optimize the digestion process.

The optimum digest mixture is species-dependent, but generally contains cellulase (1–3% w/v) and pectinase (0.1–0.5%, w/v) in 0.5 M sorbitol and 1 mM $CaCl_2$, 5 mM MES–HCl, pH 6.0. This is frequently supplemented with a range of additives (see [11]). Prior to this, the leaf material is placed in a hyperosmotic 'plasmolysing' solution (normally the digest mixture minus the enzymes) which allows the plasma membrane to 'shrink away' from the cell wall to permit efficient protoplast formation in the subsequent digest. See Box 3.5 for a typical procedure.

Box 3.5

Aim: Preparation and homogenization of plant protoplasts.

Protocol

Step 1. Incubate leaves in the plasmolysing solution at 20°C for 30 min. before cutting them into very fine pieces (0.5–1 mm) with razor blades.

Step 2. Incubate the cut pieces with the enzyme digest mixture for up to 3 h at 20°C while being illuminated.

Step 3. After filtration through nylon mesh, recover the protoplasts from the digest mixture by low speed centrifugation at 100–150*g* for 3–5 min.

Step 4. Disrupt the protoplasts by forcing a suspension from a syringe through a 20–50 μm mesh in the lysis medium (at 4°C) which is tailored to the individual plant species but normally contains approx. 0.3 M sorbitol, 2 mM EDTA and 50 mM of an organic buffer, maybe with a protective agent such as polyvinylpyrrolidone (1% w/v). The high buffer strength is required to neutralize any acid released from plant vacuoles broken during the homogenization process [11].

General applicability. Young (4–5 day old) barley or wheat are the best sources.

3.1.9 Yeast

Most of the older techniques which were used to disrupt large amounts of material used some form of mechanical shear to disrupt the tough outer wall of the organism (e.g. vigorous shaking with glass beads, ball mills etc.). Modern methods employ enzymes to digest the wall material in order to produce spheroplasts which, like plant protoplasts, can then be disrupted using liquid shear or osmotic lysis. The enzyme used is zymolase 100-T (0.25 mg ml^{-1}) in 1.4 M sorbitol, 50 mM phosphate buffer, pH 7.5. After incubation at 37°C for 45 min the spheroplasts are harvested by centrifugation and homogenized in 0.8 M sorbitol, 1 mM EDTA in an organic buffer (often triethanolamine–HCl, pH 7.2) in a Dounce homogenizer. For more details see [1] and [12].

3.1.10 Bacteria

The choice of homogenization method is determined to some extent by the type of bacterium. Broadly there are two approaches to the disruption of the tough outer wall of the organism. Older shear methods use some sort of press in which the bacterial suspension is forced through an orifice, either as a liquid (French press) or as a frozen block (Hughes press). The principle of the French press is not unlike that of the Stansted cell disrupter (see Section 3.1.1). Because of the rather cumbersome nature of the equipment, presses have given way to the use of sonication as the principal shear technique (see Box 3.6).

- Using the French press or sonication, the plasmalemma forms inside-out vesicles.
- Using the Hughes press, the plasmalemma forms large sheets.

Another approach is to disrupt the outer layers either chemically or with an enzyme. Although penicillin, for example, can be used to inhibit the synthesis of new peptidoglycan so that the bacteria outgrow their 'coat', it is more common to digest this outer layer enzymically with lysozyme (Box 3.7). Once the outer peptidoglycan layer has been removed, the protoplast (Gram-positive bacteria) or spheroplast (Gram-negative bacteria) becomes osmotically sensitive, so some sort of osmotic balancer such as sucrose is included to render the suspension slightly hyperosmotic. For more details see [13].

Box 3.6

Aim: Preparation of bacterial membranes by sonication.

Protocol

Step 1. After harvesting, wash and suspend the bacteria in a suitable medium. A typical medium is 50 mM Tris–HCl, 2 mM $MgCl_2$, 1 mM EGTA, pH 7.4 [14]. The volume of buffer is variable, but can be as little as 1.5× the volume of the bacterial pellet.

Step 2. Sonicate the sample surrounded with ice, with the tip of the probe (normally about 1 cm in diameter) a few millimeters into the sample. Always carry out sonication in short 'bursts' of approx. 15 sec interspersed with 'rests' of the same time to allow the sample to cool. Routinely 10 'bursts' at 150–200 W should be satisfactory.

Note: DNase I (10 µg/ml) in the medium will hydrolyze released DNA.

General applicability: All bacteria.

The spheroplasts themselves can be used for functional or structural studies, or they can be lysed osmotically and the plasmalemma fragmented to form vesicles (right-side out) using a hypotonic buffer (usually in a 10 mM phosphate buffer, containing 2 mM $MgSO_4$, with DNase and RNase to hydrolyze nucleic acids). If the DNA and RNA released during the formation of membrane vesicles were not hydrolyzed, the viscosity of the solution would affect subsequent harvesting by centrifugation. See [13] for more details.

Box 3.7

Aim: Preparation of bacterial membranes using lysozyme.

Protocol

Step 1. Prepare spheroplasts from an exponentially growing Gram-negative bacteria (e.g. *E. coli*) by suspending the cells in buffered 20% (w/v) sucrose containing chloramphenicol (50 µg ml^{-1}).

Step 2. Lyse the coat by stirring with lysozyme (0.5 mg ml^{-1}) and K-EDTA (10 mM) at 20°C for 30 min.

Step 3. Harvest the spheroplasts by centrifugation at 13 000g for 30 min.

Practical note
Membrane vesicles are often prepared directly from Gram-positive bacteria by carrying out the lysozyme digestion in a hypo-osmotic Mg^{2+}-containing phosphate buffer, again with DNase and RNase.

3.2 Fractionation of subcellular organelles

Fractionation techniques depend on differences between the various particles in four properties: size, density, charge and composition. There are three principal modes of fractionation: centrifugation, continuous-flow electrophoresis and immunoadsorption.

Only prokaryotes (bacteria) and the human erythrocyte contain a single type of membrane structure (the plasmalemma or plasma membrane); once these cells have been disrupted only a single centrifugation step is required to harvest the membranes. All other eukaryotic cell homogenates contain a huge range of membranous particles (see Chapter 2) which require fractionation and the most common way of doing this uses centrifugation.

Generally, the homogenate is separated crudely into three or four fractions by differential centrifugation, followed by further resolution of one of these fractions in some form of density gradient (see [14] for detailed information about centrifugation theory and practical aspects of centrifugation).

3.2.1 Differential centrifugation

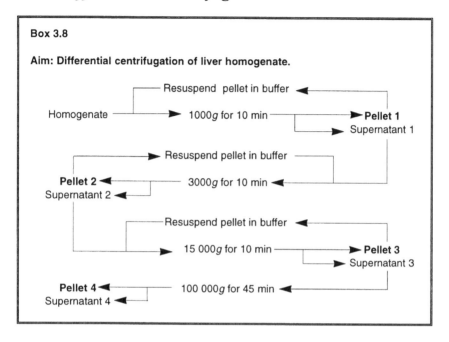

Box 3.8

Aim: Differential centrifugation of liver homogenate.

A typical differential centrifugation scheme is given in Box 3.8; the homogenate is first centrifuged at 1000*g*, and the supernatant is decanted or aspirated and recentrifuged at a higher *g*-force and/or time. This procedure is then repeated as many times as required. Routinely four of these centrifugations are carried out, as shown in Box 3.8. The composition of the four pellets from a mammalian tissue homogenate, such as liver, is given in *Table 3.3*. Broadly speaking the fractionation is achieved on the basis of particle size (see *Table 3.4*). Resolution and recoveries can be improved by washing the pellets (i.e. resuspending the pellets in medium and recentrifuging at the same speed (see Box 3.8). This is, however, time-consuming and repeated centrifugations and resuspensions lead to progressive disruption of membrane-bound particles.

- To isolate sheets of plasma membrane from a liver homogenate, a preliminary centrifugation at 280*g* for 5 min is introduced. This sediments most of the nuclei into a very loose pellet from which the supernatant must be removed very carefully. The plasma membrane sheets are then pelleted from the supernatant at 1500*g* for 10 min [15].
- Inclusion of DNase in the homogenate may be beneficial if the nuclei tend to break down and release DNA.

• Chloroplasts are dense organelles that sediment rapidly. They will sediment at 500–6000*g* for 0.5–1.0 min [10], the precise time depending on the source.

Table 3.3. Composition of differential centrifugation pellets (see Box 3.8)

Pellet number	Composition	Comment
1	Nuclei, unbroken cells, heavy mitochondria	Also plasma membrane sheets if present
2	Heavy mitochondria	
3	Mitochondria, lysosomes, heavy mitochondria	Also Golgi membranes if present as large tubules
4	Microsomes (membrane vesicles) formed from plasma membrane, endoplasmic reticulum and Golgi, plus vesicles involved in endocytosis and secretion	

Table 3.4. Dimensions of subcellular particles

Organelle	Size (μm)	Organelle	Size (μm)
Nucleus	4–12	Chloroplasts	2–5
Mitochondria	0.4–2.5	Golgi tubules	1.0–2.0
Lysosomes	0.4–0.8	Membrane sheets	3–20
Peroxisomes	0.4–0.8	Vesicles	0.05–0.5

In a few instances differential centrifugation can be used as the sole means of purification. Heavy mitochondria (3000*g* for 10 min) are routinely prepared by this method from a post-nuclear supernatant (see Box 3.15); indeed it is the method of choice for the production of mitochondria with a high respiratory quotient. Nuclei from animal cells are occasionally prepared by simple differential centrifugation although removal of contaminants by washing is not very effective. A total microsome fraction is also simple to produce from the light mitochondrial supernatant.

3.2.2 Density gradient centrifugation

For a more detailed discussion of the principles and practical considerations on the making and running of density gradients the reader is referred to [14]. Suffice here to state that density gradients can be used to separate membrane particles on the basis of their size or of their buoyant density. The former is less common as a general procedure because of the size heterogeneity of individual populations and it is also very restrictive on the amount of material which can be handled on a gradient (it has to be placed on top in a very narrow band). In buoyant density fractionation the sample size is less restrictive and can be placed at the top, at the bottom or throughout the gradient.

Pre-formed gradients (i.e. those made by the operator prior to loading the sample and centrifugation) can be discontinuous (a series of layered solutions of increasing density) or continuous (formed using a two-chamber gradient maker or by diffusion of a discontinuous one). With certain gradient media (Percoll or iodixanol) they may be self-generated, i.e. the sample is mixed with medium and during the centrifugation the gradient is formed and the particles move to their banding positions (see [14] and [16]–[18] for details).

The simplest discontinuous system is a density barrier. Density barriers can be used to isolate either the most dense or the least dense particles from a resuspended differential centrifugation fraction. In the former case the sample is layered over a solution whose density is just lower than that of the particle to be isolated; in the latter the sample is layered over a solution whose density is just higher than that of the particle to be isolated.

Gradient media. The principal media and some of their properties are listed in *Table 3.5*. Sucrose was the first medium to be used for making gradients: because it has been in use since the middle of this century, there is a huge literature on its use. The more modern media (metrizamide, Nycodenz, iodixanol and Percoll), however, although more expensive, are far more suitable for osmotically sensitive organelles.

The density of subcellular organelles in three gradient media is given in *Table 3.6*. Loss of water from the osmotically active organelles and membrane vesicles in hyperosmotic sucrose solutions increases the density of organelles and vesicles and is probably responsible for the rather similar and overlapping densities of many of these particles in this medium. In iso-osmotic gradient media such as iodixanol or Percoll, organelles retain their enclosed water, and consequently their density is generally much lower.

Table 3.5. Density gradient media for membranes

Medium	Maximum density (g ml^{-1})[a]	Osmolarity (mOsm)[b]	Viscosity (mPas)
Sucrose	1.32	>800 @ 1.10 g ml^{-1}	14 @ 1.22 g ml^{-1}
Ficoll	1.14	150 @ 1.12 g ml^{-1c}	>25 @ 1.12 g ml^{-1}
Percoll	approx. 1.25[d]	10 at all densities	5 @ 1.22 g ml^{-1e}
Nycodenz	1.32	600 @ 1.32 g ml^{-1}	6 @ 1.32 g ml^{-1}
Iodixanol	1.32	260 @ 1.32 g ml^{-1}	10 @ 1.32 g ml^{-1}

[a]Maximum density normally used in gradients.
[b]The osmolarity of mammalian fluids is 285–300 mOsm.
[c]Osmolarity increases rapidly above this density.
[d]Commercial Percoll has a density of 1.13 g ml^{-1}, with higher densities only obtained during centrifugation.
[e]Viscosity increases rapidly above this density.

Table 3.6. Density of subcellular organelles

Organelle	Density (g ml^{-1})		
	Sucrose	Iodixanol	Percoll
Nuclei	>1.32	1.22–1.24	n.a.
Peroxisomes	1.19–1.23	1.18–1.21	1.04–1.07
Mitochondria	1.18–1.21	1.14–1.16	1.07–1.08
Lysosomes	1.17–1.20	1.10–1.14	1.09–1.11
Golgi membranes	1.05–1.12	1.03–1.08	1.03–1.05
SER	1.06–1.15	1.07–1.10	1.03–1.06
RER	1.18–1.26	1.12–1.14	n.a.
Plasma membrane	1.07–1.19	1.05–1.15	1.03–1.04

n.a., data not available.

3.2.3 Density perturbation

Sometimes the density difference between two membranes is too small to be practically useful for their separation: this is particularly true of many smooth membrane vesicles. In this case the density of one of the membrane vesicle types may be perturbed to create a density difference artificially. This general approach utilizes the presence of some membrane-specific component to bind a probe molecule and thus render the membrane more dense and separable by gradient centrifugation from other membranes not bearing this marker.

For example, during homogenization, plasma membrane vesicles normally form with the *in vivo* orientation (i.e. the carbohydrate groups of their glycoproteins and proteoglycans are outermost). Lectins such as wheat-germ agglutinin (see Section 4.3.3 for the carbohydrate specificity of lectins) can bind to certain carbohydrate sequences on the surface. Membrane vesicles derived from cytoplasmic membranes have their carbohydrate groups exposed only at the cisternal face and are thus inaccessible to the lectin. The lectin is made very dense by linking it first to BSA (usually with glutaraldehyde) and then to colloidal gold [19]. Box 3.9 describes the procedure.

Box 3.9

AIm: Density perturbation of the plasma membrane.

Principle
Incubate the intact cells with the lectin–BSA–gold complex at 4°C. This temperature prevents endocytosis of the perturbant.

After the cells have been washed to remove excess lectin, the cells are homogenized and then fractionated on a sucrose gradient.

Practical note: After washing away the excess ligand, the cells could be incubated at 37°C so that the lectin is internalized: in that way endosomes can be density perturbed.

Another plasma membrane specific marker is the lipid, cholesterol. Although present in most animal cell membranes, its concentration in the plasma membrane is far higher than any other membrane: only the Golgi membrane shows a significant amount of this lipid. Digitonin is a steroid derivative (cardiac glycoside) which binds cholesterol. When added to a crude plasma membrane preparation, it raises the density of this membrane by about 0.03 g ml^{-1}, while other membranes (except for the Golgi) are not affected [20].

To density perturb endosomes the **horseradish peroxidase–3,3′–diaminobenzidine (HRP–DAB)–H$_2$O$_2$ method** is probably the most commonly used strategy. The ligand of interest is radiolabeled and coupled to HRP; it is then internalized by the cells and processed as described in Box 3.10 [21].

3.2.4 Fractionation by immunoaffinity

This technique is not unlike the density perturbation method, in that it makes use of the presence (on the surface of the membrane) of antigens which can bind noncovalently to a probe molecule (ligand) – often an antibody raised to that determinant. The antibody is then bound to a secondary antibody or Protein A, linked to some form of bead. It is thus very similar to affinity chromatography except that the beads which bear the antibody are generally not formed into a column

Box 3.10

Aim: Density perturbation using HRP and DAB to fractionate endosomes.

Protocol

Step 1. Bind radiolabeled ligand linked to HRP to cell surface.

Step 2. Internalize ligand.

Step 3. Homogenize cells and prepare microsomes.

Step 4. Control: incubate with DAB (no density shift).

Step 5. Test: incubate with DAB and H_2O_2; DAB polymerizes (density shift).

Interpretation: The microsomes are analyzed on a density gradient (1.1–1.13 g ml^{-1} at 100 000g for 3–4 h). The position of the endosome fraction is identified by the radiolabel.

When the control and test incubations are compared, only the peroxide-treated test microsomes show an increased density banding of the radiolabel.

Applicability: Under optimal labeling, the presence of oxidized DAB shifts the density by up to 0.06–0.08 g ml^{-1}. The preparation of the DAB has to be carefully monitored and amounts of the HRP-ligand, DAB etc. may have to be adjusted to suit a particular sample [21].

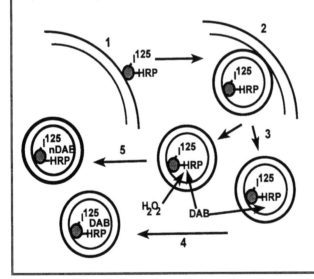

but rather simply mixed with a membrane suspension and the membrane–antibody–linker-bead complex pelleted by low-speed centrifugation. The method has been developed by a number of workers [22, 23] and the variations are in the type of bead (in some instances a magnetic bead has been used, so the bound membranes are harvested in a magnetic rather than a centrifugal field) and in the

---

Box 3.11

Aim: Isolation of sinusoidal membrane of liver plasma membrane by immunoaffinity.

Protocol

Step 1. Antibody to extracellular domain of SC is linked to Sepharose–Protein A (S–PrA).

Step 2. Complex washed in detergent-containing buffers.

Step 3. Complex incubated with microsomal fraction from rat liver; only those vesicles bearing the SC extracellular domain at their surface bind to the S–PrA antibody.

Step 4. Centrifugation in a microfuge sediments the bead-bound vesicles, which are then washed several times in a suitable buffer.

Practical note: A control incubation with a nonimmune serum is always included.

General applicability: Applicable to any vesicles which bear a specific protein domain at their surface to which an antibody can be raised, e.g. secretory vesicles bearing the cytoplasmic domain of SC on their outer surface.

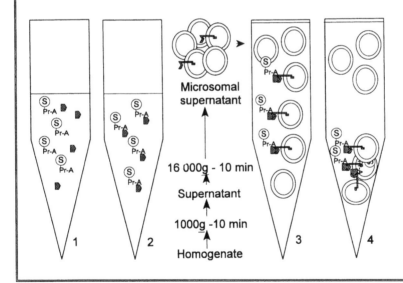

---

length of linker molecule. The linker molecule improves the binding of the antibody to the beads by removing some steric hindrance.

Generally it is better to bind the membrane-specific antibody to the membrane before reacting with the linker-bead rather than bind the membrane to a specific antibody–linker-bead complex, although the latter has been successfully used. A nonimmune serum is always used as a control to judge the specificity of the binding. Some of the applications are described in *Table 3.7* and, in the example given (Box 3.11), an antibody to the polymeric IgA receptor (often called the secretory component, SC) which binds specifically to the sinusoidal domain of the liver plasma membrane is used to isolate that membrane domain [23].

Table 3.7. Immunoaffinity isolation of membranes

Bead	Linker	Application	Reference
Diazocellulose	Anti-IgG antibody	Cholinergic nerve terminals	22
Sepharose	Protein A	Sinusoidal domain (liver plasma membrane)	23
Staphylococcus aureus[a]	Protein A	Cytoplasmic vesicles	24

[a] The bacterium itself which bears Protein A acts as the bead.

3.2.5 Fractionation by continuous flow electrophoresis

In this method membranes are separated on the basis of their surface charge density. It has never achieved great popularity, in spite of the fact that fractionations can be achieved under very mild conditions in iso-osmotic media of low density and viscosity and that the separation is virtually independent of sample size. The equipment is, however, expensive and rather few applications have been developed.

Simply, the apparatus consists of a vertical refrigerated chamber made from two glass plates approx. 60×10 cm separated by approx. 0.7 mm. Down this chamber flows a continuous stream of buffer (0.25 M sucrose in 10 mM triethanolamine–10 mM acetic acid, pH 7.4, containing 1 mM EDTA is the most commonly used medium), between the two electrodes. The sample is injected into this moving curtain of buffer: particles in the sample which are uncharged will move directly downwards from the injection port (see *Figure 3.6*) but charged particles will be deflected to an extent which is proportional to their surface charge density. At the bottom of the chamber the effluent is divided by a manifold into 90 outlet tubes. Typically the applied voltage is 1100–1750 V and the buffer flow rate is 300–500 ml h^{-1}. These two parameters determine the extent of the deflection of charged particles across the chamber. The lower the buffer flow rate and the higher the imposed voltage, the greater will be the deflection.

It might be expected that the principal application would be the separation of plasma membrane since, under normal conditions, this membrane has a high surface negative charge because of the presence of sialic acid residues; however, this is not the case. It has been used to fractionate the cytoplasmic membranes of platelets, but only after

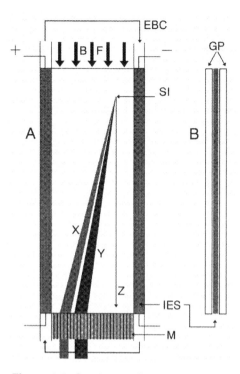

Figure 3.6. Continuous flow electrophoresis: (a) face view; (b) side view. The electrodes (+ve on left) are bathed in a circulating flow of electrode buffer (EBC) and electrical continuity with the buffer (BF), which flows continuously down the chamber, is made by ion exchange strips (IES). The sample is introduced into the moving buffer at the injection port (SI). Separation of two types of particle with high (X) and medium (Y) electronegativity is shown; any uncharged particles (Z) would migrate vertically in the buffer flow (BF). The manifold (M) at the bottom of the chamber divides the effluent of the chamber into 90 fractions. Buffer flow is maintained by a peristaltic pump beyond the manifold.

treatment of the cells with neuraminidase [25]. The partial separation of different Golgi compartments has also been achieved using this technique [26]. Most organelles, however, appear to possess rather similar surface charge densities and they are consequently not separable by this technique.

One of the more successful applications is the resolution of lysosomes and endosomes [27], but even here treatment of the sample with N-tosyl-L-phenylalaninechloromethylketone (TPCK)-trypsin is necessary to provide the best resolution. As for all separative techniques, the particles in the sample must be free from aggregation, and this is particularly important for continuous flow electrophoresis as aggregates may seriously disturb the flow path of the particles and may even block the entrance port. For this reason it is recommended that repeated low speed pelleting of the material is carried out prior to injection.

3.2.6 Fractionation in aqueous two-phase systems

When a solution of dextran (M_r 500 000) is shaken with a solution of polyethyleneglycol (PEG) 3350 (under defined conditions of concentration, pH and salt concentration), they separate into two phases, with PEG on the top. If this is carried out in the presence of a mixture of membranes then some membranes partition preferentially into the upper phase and some preferentially into the lower phase. The distribution of a particular membrane between the two phases is determined principally by its surface charge and its hydrophobicity.

As with all phase partition separations the standard technique is to harvest the two phases and then repeat the process as many times as necessary to achieve a satisfactory enrichment of the membrane of interest: the lower dextran layer is repartitioned with fresh upper layer and the upper PEG layer repartitioned with fresh lower layer. It is possible to automate this procedure to allow as many as 20–30 sequential partitionings, but for normal purposes, three or four times is adequate. Box 3.12 gives a typical protocol (from [3]).

Box 3.12

Aim: To separate membranes by two-phase partitioning.

Protocol

Step 1. Thoroughly mix 200 g of 20% dextran with 103 g of 30% PEG, 33 ml of 0.2 M phosphate buffer, pH 6.5, and 179 ml water.

Step 2. Allow the two layers to separate, either overnight or by centrifugation: these serve as upper and lower layers for the repeated partitions.

Step 3. Shake a crude membrane pellet with the dextran/PEG mixture (30–40 inversions); centrifuge at 2000g for 10 min and separate.

Step 4. Shake the lower dextran layer with fresh upper phase and the upper PEG layer with fresh lower phase.

Applicability: Plasma membranes seem to have the highest affinity for the upper PEG layer and endoplasmic reticulum the lowest, with mitochondria and other organelles in between. The technique is normally only used for the isolation of plasma membrane.

3.3 Identification of fractionated membranes

Once the membranes from a tissue or cell have been fractionated by one or more of the methods described above, they have to be identified. This is of course not a problem where the method of isolation is based on the presence of a specific marker in the membrane of interest to which an antibody is available. Otherwise there are three basic types of identification used in routine fractionation of tissues and cells: morphological, chemical and enzymic.

3.3.1 Morphological

Phase-contrast microscopy is routinely used to check the progress of homogenization of a cell or tissue, but the only two membrane particles which can be identified unambiguously are nuclei and large sheets of plasma membrane (e.g. the contiguous membrane domains derived from hepatocytes or intestinal brush border). Phase-contrast microscopy also provides some indication of the integrity of the nuclei, which should be the same size as the nuclei in the intact tissue and appear dark gray.

The transmission electron microscope can identify all of the other major organelles, mitochondria, lysosomes, peroxisomes, Golgi apparatus, rough endoplasmic reticulum (RER) and membrane vesicles. But to discriminate the different types of smooth membrane vesicles derived from the SER, plasma membrane, Golgi, endosomes, secretory vesicles etc. is rather less easy, unless they can be identified by size or some associated structure. Secretory vesicles often have extremely electron-dense contents (the so-called granular fraction of secretory cells) and plasma membrane vesicles may be associated with some cell coat (glycocalyx) material on their outer surface. When membrane fractionation has been used in conjuction with the use of an endocytosed ligand which has been tagged with an electron-dense particle (e.g. colloidal gold), then electron microscopy can identify vesicles containing the marker.

3.3.2 Chemical

Most routine chemical assays rely on spectrophotometry to measure the intensity of some colored product. It is thus impossible to use fractions in Percoll directly in such assays because of its light scattering, and iodinated density gradient media absorb significantly below 340 nm. So Percoll must always be precipitated prior to any spectrophotometric assay and the iodinated density gradient media must be removed by dilution of the membrane fractions with an equal volume of buffer and sedimenting of the membrane at some suitable g-force if the required wavelength is below 340 nm (in practice this is rarely required). Some of the more widely used protocols are contained in [30].

Nuclei. Although nuclei can be identified and quantitated by phase-contrast microscopy, it is often more desirable to estimate the DNA in the various fractions produced by a fractionation protocol. There are two types of assay.

- The standard spectrophotometric assay for DNA uses a solution of diphenylamine to measure the deoxyribose sugar. It cannot, however, be performed directly on samples in sucrose, Ficoll or metrizamide (all carbohydrate-containing gradient media), without first precipitating the DNA with 0.25 M perchloric acid. Membrane fractions in Nycodenz or iodixanol, however, can be assayed directly.
- More sensitive assays include the 4',6'-diamidino-2-phenylindole (DAPI) assay. This has the merit of simplicity, relying on the enhanced fluorescence of the molecule (approx. 20 times) in the presence of DNA.

RER. The RNA in the ribosomes of the RER can be measured by standard spectrophotometric assays or by a more sensitive fluorescent assay.

- There are two chemical assays: one measures the ribose content (orcinol assay) and the same restrictions apply as to the diphenylamine assay for DNA, i.e. any gradient media containing sugar must be removed by perchloric acid precipitation of the RNA; the alternative method uses methyl green and is not affected by the presence of sugar.
- The fluorescent assay uses ethidium bromide; it is very sensitive and is not affected by the presence of gradient media.

Plasma membrane. There are three components of the plasma membrane which can act potentially as markers: cholesterol, sialic (N-acetylneuraminic) acid and sphingomyelin. However, none of these is used routinely to monitor a fractionation procedure.

Chloroplasts. The chlorophyll content of these organelles can easily be measured simply by extraction with acetone, precipitation of the residual insoluble material at 3000*g* and measuring the absorbance at 652 nm.

3.3.3 Enzyme markers

The standard markers are enzymes whose presence in certain membranes has been determined either by cytochemical procedures on intact tissue, by known physiological functions in the intact tissue or by association with membranes which can be identified morphologically in the light or electron microscope.

An important consideration is the possible effect of the separation medium on enzyme activity. Generally, inhibitory effects of gradient media on enzyme activity are relatively small and observable only at higher concentrations. Moreover they are usually reversible and the concentration of the medium in the assay is normally significantly diluted by the assay mixture below the concentration where either an inhibitory or enhancing effect is observed. The same caveats apply here as to the chemical assays, i.e. Percoll suspensions cannot be used directly in spectrophotometric assays, nor can iodinated density gradient media at wavelengths below 340 nm.

In many cases a gradient can be analyzed directly, so long as there is sufficient material in the sampled volume to be detectable by the assay. Otherwise it is necessary to concentrate the fractions by sedimenting the membranes and resuspending them in a smaller volume of homogenization medium or analysis buffer. If this is carried out for one fraction it should be performed on all fractions.

- Ideally, volumes of each fraction should be taken which contain the same amount of protein and sufficient for a single assay, and these diluted with buffer and pelleted in microcentrifuge tubes which can be used subsequently for the assay.
- It is important that the enzyme (membrane) concentration should be the rate-limiting factor – *not* the substrate. It is, therefore, always a good idea to measure the rate of reaction at two time points and at two protein concentrations, at least. The activity should be linear with time and protein concentration; the specific activity, expressed as units min^{-1} mg^{-1} protein, should be the same and independent of the time and protein concentration.

Some assays can be performed by the continuous monitoring of the absorbance of a product in a recording spectrophotometer, in which case a linear response over a measurable time span is the aim. In other cases where the product is measured by addition of a secondary reagent which effectively stops the reaction, then it is often convenient to set up a large enough assay so that samples can be withdrawn at set times into tubes containing the secondary reagent.

It is not the intention of this book to provide detailed protocols for each enzyme (these are available in several texts [29,30]), but some comments will be made on the ease (or otherwise) and applicability of certain assays.

Mitochondria. Malate dehydrogenase and succinate dehydrogenase are the most commonly used markers.

Malate dehydrogenase activity is measured by the oxidation of NADH in the presence of oxaloacetate, via the decrease in absorbance of the NADH at 340 nm. Rotenone and Na$_2$S are often included to prevent oxidation of NADH by the electron transport chain and NADH oxidase.

Succinate dehydrogenase activity is measured either by reduction of cytochrome c in the presence of KCN to prevent reoxidation of the reduced cytochrome c by cytochrome oxidase, or by reduction of an artificial electron acceptor, *p*-iodonitrotetrazolium violet. Either of these methods is more convenient than measuring malate dehydrogenase since they both use wavelengths in the visible part of the spectrum.

- Mitochondria from tissues such as beef heart and rat liver form a major subcellular fraction and the enzymes are easily measurable. Cultured animal cells, however, derive a lot of their energy requirements from anaerobic glycolysis and levels of mitochondrial enzymes tend to be rather low.

Lysosomes. The most widely used marker enzymes for these organelles are acid phosphatase, β-galactosidase and β-N-acetylglucosaminidase. Their pH optima are all in the range 4–5.

Substrates for these enzymes are commercially available as *p*-nitrophenyl derivatives so that they are easily measured by the release of nitrophenol which has a bright yellow color in alkaline solution (absorbance measured at 405–410 nm).

The actual pH used to measure the nitrophenol depends to some extent on the substrate. *p*-Nitrophenylphosphate, the substrate for acid phosphatase, is stable even at very high pH values; some substrates are less stable and use a pH 10 glycine–NaOH buffer to stop the reaction.

Peroxisomes

Catalase breaks down H_2O_2 to oxygen and water. To prevent nonenzymic breakdown of H_2O_2 the assay is carried out at 0°C. The rate of reaction is measured by determining the amount of substrate remaining after action of the enzyme (i.e. a back titration). There are two main methods of measuring the residual H_2O_2; one by the decoloration of $KMnO_4$ and the other by reaction with titanium oxysulfate. The latter is far easier, a stable yellow product being produced under acid conditions.

Golgi membranes. The Golgi is the site of the terminal glycosylation of glycoproteins and any of the enzymes associated with this activity can be used as a marker. Galactosyl transferase (UDP-galactose galactosyl transferase) has been used as a general Golgi marker and the acceptor used for the galactose is either N-acetylglucosamine or a suitable glycoprotein, e.g. desialylated, degalactosylated fetuin or ovalbumin. All methods use UDP-galactose, radiolabeled in the sugar moiety.

The product of the transfer of galactose to N-acetylglucosamine, N-acetyllactosamine, is separated from UDP-galactose on an anion exchange resin. Although the separation merely involves applying the assay mixture to a small column in a Pasteur pipette and eluting the column with water into a counting vial, the processing of many samples is extremely tedious.

Using ovalbumin an aliquot of the incubation mixture is spotted on to a filter disc, allowed to dry and the protein precipitated on the disc by immersion in trichloroacetic acid; the labeled UDP-galactose is soluble and removed by washing.

Galactosyl transferase is actually a marker for the *trans*-Golgi, but as long as the Golgi remains in an intact state during isolation, the enzyme can be used as a general marker. A marker for the *cis* domain is N-acetylglucosaminylphosphotransferase and for the median domain, N-acetylglucosaminyl transferase I.

Endoplasmic reticulum. In liver, either glucose-6-phosphatase or NADPH–cytochrome c reductase are used as markers for this membrane. With the exception of kidney, however, the levels of glucose-6-phosphatase in other tissues or in cultured cells are very low; indeed there is very little physiological need for this enzyme in any tissue which does not regulate blood glucose.

Glucose-6-phosphatase is measured by estimating the release of inorganic phosphate from the substrate. It is a very simple procedure but, as with all phosphate measurements, great care has to be taken to eliminate phosphate from other sources by acid-washing all glassware and any plasticware should be phosphate-free.

There are a vast number of phosphate colorimetric assay procedures, most of which involve complexing the phosphate with ammonium molybdate, followed by reaction with either 1-amino-2-naphthol-4-sulfonic acid, diaminophenol or ascorbic acid [29]. Generally amounts of phosphate in the range 60–300 nmol are adequate but there are a variety of modifications to make the assay more sensitive (10–100 nmol).

NADPH–cytochrome c reductase is present in most tissues and cells and the assay is easy to carry out using a recording spectrophotometer.

Plasma membrane. The data on liver and other organized tissues are far more clear than on cultured cells. Not only are their surface membranes characterized by high levels of certain enzymes, there is also a considerable body of evidence concerning domain specificity. Generally speaking the apical domain is associated with high levels of 5′-nucleotidase (an enzyme which hydrolyzes AMP to free adenosine and phosphate), alkaline phosphatase, alkaline phosphodiesterase and leucine aminopeptidase. The basolateral membrane is normally characterized by high levels of the Na^+/K^+-ATPase. In liver the latter enzyme is predominantly a contiguous membrane enzyme while the sinusoidal membrane is enriched in the glucagon-stimulated adenylate cyclase. Enzyme polarity is also shown by some differentiated cultured cells (e.g. Caco 2 cells grown on filter supports).

5′-Nucleotidase is most commonly determined by measuring the release of inorganic phosphate from unlabeled AMP. Alternatively it can be measured using AMP radiolabeled with ^{14}C or ^3H in the ribose moiety; unreacted labeled substrate is removed either on an anion exchange column or by adsorption to $Ba(SO)_4$.

AMP is also hydrolyzed by the alkaline phosphatase which is present in the plasma membrane. This can be overcome, in a radioassay, by including an excess of β-glycerolphosphate as a substrate for the phosphatase enzyme (the 5′-nucleotidase is specific for AMP – and some other nucleotides). Alternatively, levamisole can be used as an inhibitor of the alkaline phosphatase.

Rapidly growing cells have particularly low concentrations of 5′-nucleotidase.

Alkaline phosphatase and alkaline phosphodiesterase are simple to assay as *p*-nitrophenol derivatives of substrates are available (*p*-nitrophenyl phosphate and thymidine-5′-monophosphate-*p*-nitrophenyl ester, respectively).

Unlike the acid phosphatase, the alkaline phosphatase is Mg^{2+}-stimulated.

The most commonly used substrate for leucine aminopeptidase is a *p*-nitrophenol derivative, leucine-*p*-nitroanilide.

There are alternative substrates such as leucyl-β-naphthylamide, which were in common use about 20 years ago, but they and their products are potent carcinogens and *should not be used*.

Na^+/K^+-ATPase is one of many enzymes in the cell which will hydrolyze Mg-ATP. The easiest way of measuring the enzyme, which is specifically stimulated by Na^+/K^+, is to measure the total ATPase activity in the presence of Mg^{2+}, Na^+, and K^+, in the presence and absence of the Na^+/K^+-ATPase inhibitor, ouabain (approx. 1 mM).

ATP hydrolysis is normally measured by determining inorganic phosphate, but an alternative assay which can be very sensitive is to use [γ-^{32}P]-ATP, taking advantage of the fact that ATP can be removed on activated charcoal and the $^{32}PO_4^{3-}$ released by hydrolysis can be counted in water by Cerenkov radiation. Before using ^{32}P-labeled material, you must consult your Safety Officer about safe handling and disposal.

The Na^+/K^+-ATPase of cultured cells tends to be rather insensitive to ouabain.

3.3.4 Immunological markers

With the advent of new techniques in DNA and protein analysis, together with the availability of monoclonal antibodies to proteins and their ready detection by immunoblotting after separation on polyacrylamide gels (see Chapter 4), a whole new library of potentially useful markers for various membranes and membrane compartments is being developed. It is beyond the scope of this chapter to cover this subject, since many of them are very specific to the particular function or membrane event under study.

3.3.4 Expression of data

It is important that once the enzymes, proteins or other chemical markers have been satisfactorily measured that the data be expressed in a useful form. The aims of preparative and analytical fractionation are rather different. In analytical fractionation, the distribution of some functional parameter across a gradient is either compared with that of another parameter or compared to that observed under different experimental conditions. A preparative fractionation on the other hand requires a careful assessment of the purity and yield of the membrane fractions.

Two parameters are important; the recovery of the membrane(s) of interest and their purity.

• The recovery of a particular membrane in each fraction is routinely expressed as the amount of marker in that fraction as a percentage of that in the total homogenate; or sometimes as a percentage of that which was applied to a gradient.
• The concentration of a membrane in a particular fraction is expressed as the specific activity of the marker (activity per mg protein).
• The enrichment of a membrane in a particular fraction can be expressed as the relative specific activity (specific activity of the marker in the fraction divided by that in the homogenate).

A useful method of expression which provides an overall 'view' of the fractionation procedure (both in terms of enrichment of a particular marker and the amount recovered) is to plot relative specific activity against the protein content of the fraction expressed as a percentage of the total protein. An example is given in *Figure 3.7*.

Another useful concept which measures the efficacy of a gradient is that of *relative concentration*, which is the concentration of an enzyme

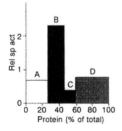

Figure 3.7. Expression of enzyme data (concentration and amount). Distribution of peroxisomal D-aspartate oxidase in four fractions: A, 700*g*/10 min pellet; B, 25 500*g*/20 min pellet; C, 170 000*g*/60 min pellet; D, 170 000*g*/60 min supernatant. Data adapted from Zaar, K., Völkl, A. and Fahimi, D. (1989) with permission from the Biochemical Society and Portland Press [31].

in a fraction/the average concentration in all the gradient fractions (assuming a uniform distribution across the gradient), which is plotted against the gradient volume (see *Figure 3.8a*).

If the gradient is not a linear one, simple measurements of specific activity or relative concentration may provide a misleading picture of a gradient fractionation. In the example in *Figure 3.8* the gradient has a median shallow region which separates two steeper regions. It appears that the gradient separates two populations of lysosomes. In

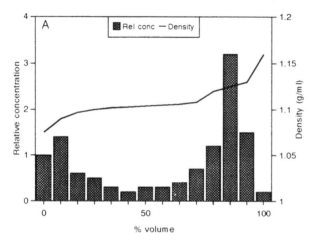

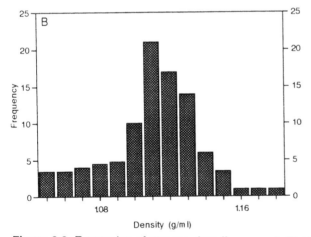

Figure 3.8. Expression of enzyme data (frequency). Distribution of N-acetyl-β-glucosaminidase in a Percoll gradient. (a) Density profile of gradient and enzyme activity expressed as a relative concentration (= actual enzyme concentration in fraction/concentration of enzyme if uniformly distributed across gradient). (b) Frequency = amount of enzyme in each fraction/density interval of fraction. Data adapted from Draye, J.-P., Courtoy, P.J., Quintart, J. and Baudhuin, P. (1987) with permission from European Journal of Biochemistry [32].

reality the enzyme profile is an artifact of the gradient shape. When the relative concentration is plotted as 'frequency' (amount of marker in a fraction/density interval of the fraction) against density, a single asymmetrical peak is observed. The gradient confirms heterogeneity of the total lysosome population but not the presence of two distinct subpopulations.

3.4 Some specimen separations of organelles and membranes

3.4.1 Nuclei

Because nuclei are the densest of all the organelles, they can be purified using a simple density barrier system (Box 3.13).

Box 3.13

Aim: To purify nuclei from a homogenate.

Protocol

Step 1. Adjust homogenate to 25% (w/v) iodixanol and layer over a barrier of 30% (w/v) iodixanol.

Step 2. Centrifuge at 10 000g for 20 min and recover the pelleted nuclei.

Principle: Only the nuclei are dense enough and large enough to sediment through the density barrier under these centrifugation conditions [17].

Applicability: The method was designed for mammalian liver; if a layer of 35% iodixanol is included the nuclei will band at the lower interface. The method can be used with a crude nuclear pellet rather than a homogenate. For cultured cells it may be necessary to change the density of the two layers. For plant cells containing starch, a crude nuclear pellet should be suspended in 35% iodixanol and layered under 30% iodixanol; the starch granules will pellet and the nuclei will float to the interface.

If a sucrose density barrier is used, the concentration of sucrose required is approx. 70% w/v (density = 1.26 g ml^{-1}); in these dense, high osmolarity solutions of sucrose nuclei lose water and consequently have a high density. Because of the high viscosity of the sucrose solution, a g-force of at least 100 000g for 1 h is required to sediment the nuclei through the barrier.

3.4.2 Golgi membranes

Golgi membranes are the least dense of the organelles and can be purified by a simple density barrier (Box 3.14). The source material is normally the light mitochondrial fraction (see Section 3.2.1) and in this case the high osmolarity of sucrose is a benefit since this causes the other organelles in this fraction (lysosomes, mitochondria etc.) to have a much higher density.

Box 3.14

Aim: To purify Golgi membranes.

Protocol

Step 1. Layer a light mitochondrial fraction (in 0.25 M sucrose) over 40% (w/v) sucrose.

Step 2. Centrifuge at 120 000*g* for 30 min. Only the Golgi membranes are of sufficiently low density to band above the interface [33].

Applicability: The method was developed for mammalian liver in which the Golgi is recovered as tubules; if the Golgi vesiculates during homogenization then a microsomal fraction may be used.

Sometimes flotation through a density barrier is used. If the light mitochondrial fraction itself is simply adjusted to 40% sucrose, Golgi membranes will float to the surface during the centrifugation.

3.4.3 Mitochondria

It is not normally necessary to use a gradient system to purify mitochondria as they can be isolated in a relatively pure form (at least in the case of mammalian tissues such as liver and heart) from a washed heavy mitochondrial fraction (see Section 3.2.1 and Box 3.15).

3.4.4 Resolving a light mitochondrial fraction

Because this fraction is a common source of a number of organelles, it is instructive to consider how a protocol for a specific organelle might be developed and this is summarized in Box 3.16.

Box 3.15

Aim: Isolation of mitochondria (highly coupled).

Protocol

Step 1. Homogenize the tissue in 0.2 M mannitol, 50 mM sucrose, 1 mM EDTA, 10 mM Hepes–NaOH, pH 7.4 (see Section 3.1.5).

Step 2. Use the first two steps of the differential centrifugation protocol (see Box 3.8) to prepare a heavy mitochondrial fraction.

Step 3. Wash the pellet at least three times to remove lighter contaminants, cytosolic proteins and fat droplets and resuspend the pellet in a suitable medium.

Applicability: The method is designed for rat liver; bovine liver and bovine heart are other common sources for functional studies on mitochondria.

Practical note
The washes in Step 3 remove, in particular, lysosomes, which contain degradative enzymes. The triglyceride must also be removed from the meniscus of the medium after each centrifugation since fatty acids are potent uncouplers of oxidative phosphorylation.

Box 3.16

Aim: Resolving an organelle from a light mitochondrial fraction (LMF).

Protocol

Step 1. Select a medium which is likely to provide the highest resolution; from *Table 3.6*, iodixanol would be a reasonable choice for fractionating the LMF.

Step 2. Prepare a linear gradient of iodixanol which covers a broad density range: 1.05–1.18 g ml^{-1} with the sample layered beneath the gradient in a density of 1.20 g ml^{-1} would be suitable.

Step 3. Centrifuge at 100 000g for 2 h in a swinging-bucket rotor and analyze the gradient for enzyme markers (see Section 3.3).

Step 4. Select a narrower range of densities and/or modulate the shape of the density profile for a gradient best suited to the purification of a particular organelle.

General applicability: *Figure 3.9a* and *b* compare two density profiles whose shallow regions can optimize the linear separation of the lysosomes or peroxisomes, respectively, from the mitochondria.

Time of centrifugation is another variable which may be modulated to optimize a particular separation.

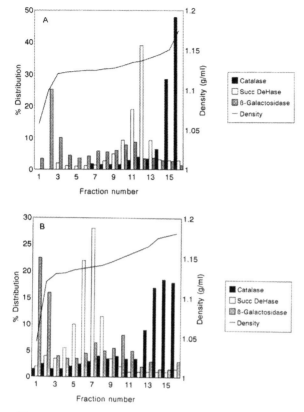

Figure 3.9. Effect of density profile on enzyme distribution. Rat liver light mitochondrial fraction (in iodixanol, ρ = 1.185 g ml^{-1}) layered beneath a 12 ml iodixanol gradient: (A) ρ = 1.12–1.16 g ml^{-1} and (B) ρ = 1.13–1.18 g ml^{-1} and centrifuged at 100 000g for 2 h in a swinging-bucket rotor. Succ DeHase = succinate dehydrogenase. The sharp drop in density at the top of the gradient is due to the overlay of 1 ml of 0.25 M sucrose.

3.4.5 Plasma membrane

Box 3.17

Aim: Isolation of plasma membrane sheets.

Protocol

Step 1. Use 500g rather than 1000g to pellet most of the nuclei from a homogenate.

Step 2. Sediment the sheets from the 500g supernantant at 1500g for 10 min.

Step 3. When the sheets are suspended in sucrose of density 1.18 g ml^{-1} and centrifuged at 113 000g, only the plasma membranes are of sufficiently low density to float to the surface [15].

Principle: The high osmotic pressure of the sucrose solution causes the osmotically sensitive contaminants (mitochondria and nuclei) to have a higher density; they therefore sediment. Plasma membrane sheets have no significant water compartment.

The plasma membrane is generally one of the more difficult membranes to prepare: only in the case of an organized tissue such as liver, which can yield large sheets of plasma membrane during gentle homogenization, is the task reasonably simple (see Box 3.17), otherwise the plasma membrane becomes just one of the many populations of membrane vesicles (see Sections 3.2.3 and 3.2.4).

3.4.6 Membrane vesicles

The complex vesicular and tubular compartments of the endocytic and secretory systems can be partially resolved using a variety of gradient systems on the basis of buoyant density or sedimentation rate. Nycodenz, iodixanol and sucrose–D_2O gradients are popular choices. The higher density of D_2O compared to H_2O means that lower concentrations of sucrose are required to cover a particular density range, and consequently the gradients have lower osmotic pressure and viscosity.

- Generally, the density of the membranes involved in secretion decreases from the rough endoplasmic reticulum (RER), to the smooth endoplasmic reticulum (SER), through *cis*-Golgi to *trans*-Golgi and the *trans*-Golgi network.
- The endocytic system has been resolved both on the basis of density and sedimentation rate. Coated vesicles, for example, are more dense than primary endosomes and some of the late endosomal compartments are larger than the earlier ones.

The vast range of published centrifugation systems have been elaborated for specific tasks, so only some broad principles will be given: a general one for fractionating vesicles on the basis of density in sucrose (Box 3.18) and a system using iodixanol for fractionating on the basis of size (Box 3.19). Both of these use the traditional approach in which the sample is layered on a pre-formed gradient. For comparison a third sytem is included (Box 3.20) which uses iodixanol as a self-generated gradient, which is much more simple to execute but requires a vertical rotor (see [16] and [18] for more information on the use and formation of self-generated gradients).

Box 3.18

Aim: Fractionation of vesicles on the basis of density in sucrose gradients.

Principles

Construct one of the following:
(1) A linear 16–70% (w/v) sucrose gradient, equivalent to a density range of 1.06–1.25 g ml^{-1}).
(2) A discontinuous gradient covering the same density range.
(3) A discontinuous gradient of 10, 12.5, 15, 17.5, 20, 22.5, 25, 27.5, 30 and 50% (w/v) sucrose in D$_2$O.

Load the sample on to (or beneath) the gradient and centrifuge at 100 000–160 000g for 3–4 h in a swinging-bucket rotor.

Applicability: These gradients have been widely used to analyze elements of the secretory system: for example an RER–Golgi intermediate [34] has been identified in the sucrose–D$_2$O system. Once the banding of a particular component has been identified, shallower gradients covering smaller density ranges can be constructed.

Box 3.19

Aim: Fractionation of vesicles on the basis of sedimentation rate (size).

Principle

Construct a continuous linear gradient of 10–40% (w/v) sucrose (1.038–1.15 g ml^{-1}) or iodixanol (density 1.04–1.20 g ml^{-1}) in a tube for a swinging-bucket rotor.

Load the sample on top of the gradient and centrifuge at approx. 80 000g for 15–40 min [35].

Applicability: Optimization of both the centrifugation time and the gradient will be required to suit the particular experimental requirements.

Box 3.20

Aim: Fractionation of vesicles on the basis of density in a self-generating gradient of iodixanol.

Principle
Adjust a microsomal fraction or a post-mitochondrial supernatant to 12.5–20% (w/v) iodixanol and centrifuge in a vertical rotor for 1–2 h at approx 350 000g_{av}.

Applicability: The chosen density and time of centrifugation will depend on the purpose of the centrifugation. A 20% gradient (2 h) provides a wide density profile for fractionating the SER and RER [36] (see *Figure 3.10*) while a 12.5% gradient (1 h) would provide a shallow gradient for separating endosomes [37].

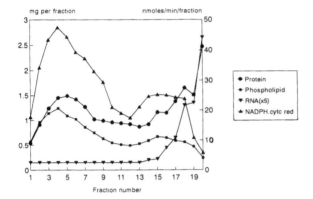

Figure 3.10. Fractionation of rat liver microsomes in a self-generated iodixanol gradient. Distribution of markers in a gradient generated from 20% (w/v) iodixanol in a Beckman VTi65.1 vertical rotor (353 000*g* for 2 h). SER distributed in top half of gradient, RER in bottom half. NADPH.cytc red = NADPH–cytochrome c reductase activity (right hand abscissa); all other markers, left-hand abscissa.

In this chapter we have tried to cover the most important aspects of membrane fractionation. Its length reflects the bewilderingly large array of methods which confronts the new research worker, but we hope that it will help in the choice of a method which is appropriate to the researcher's particular requirements. The next chapter deals with some of the basic methods of compositional analysis.

References

1. **Graham, J.M.** (1997) in *Subcellular Fractionation – A Practical Approach* (J.M. Graham and D. Rickwood, eds), pp. 1–29. IRL Press at Oxford University Press, Oxford.
2. **Balch, W.E., Dunphy, W.G., Braell, W.A. and Rothman, J.E.** (1984) *Cell* **39:** 405–416.
3. **Evans, W.H.** (1992) in *Preparative Centrifugation – A Practical Approach* (D. Rickwood, ed.), pp. 233–270. IRL Press at Oxford University Press, Oxford.
4. *Protease Inhibitors*, Calbiochem-Novabiochem Technical Sheet CB0572 (1995). Calbiochem-Novabiochem Int., San Diego, CA.
5. **Chappel, J.B. and Perry, S.V.** (1954) *Nature* **173:** 1094–1095.
6. **Marsh, M., Schmid, S., Kern, H., Harms, E., Male, P., Mellman, I. and Helenius, A.** (1987) *J. Cell Biol.* **104:** 875–886.
7. **Graham J.M.** (1993) in *Methods in Molecular Biology*, Vol. 19 (J.M. Graham and J.A. Higgins, eds), pp. 97–108. Humana Press, Totowa, NJ.

8. **Goldberg, D.E. and Kornfeld, S.** (1983) *J. Biol. Chem.* **258:** 3159–3165.
9. **Dunphy, W.G. and Rothman, J.E.** (1983) *J. Cell Biol.* **97:** 270–275.
10. **Whitehouse, D.G. and Moore, A.L.** (1993) in *Methods in Molecular Biology*, Vol. 19 (J.M. Graham and J.A. Higgins, eds), pp. 123–131. Humana Press, Totowa, NJ.
11. **Moore, A.L. and Whitehouse, D.G.** (1993) in *Subcellular Fractionation – A Practical Approach* (J.M. Graham and D. Rickwood, eds), pp. 243–270. IRL Press, Oxford.
12. **Guerin, B., Labbe, P. and Somlo, M.** (1979) *Methods Enzymol.* **55:** 149–159.
13. **Poole, R.K.** (1993) in *Methods in Molecular Biology*, Vol. 19 (J.M. Graham and J.A. Higgins, eds), pp. 109–122. Humana Press, Totowa, NJ.
14. **Ford, T.C. and Graham, J.M.** (1991) *An Introduction to Centrifugation.* BIOS Scientific Publishers, Oxford.
15. **Scott, L., Schell, M.J. and Hubbard, A.L.** (1993) in *Methods in Molecular Biology*, Vol. 19 (J.M. Graham and J.A. Higgins, eds), pp. 59–69. Humana Press, Totowa, NJ.
16. **Dobrota, M. and Hinton, R.** (1992) in *Preparative Centrifugation – A Practical Approach*, (D. Rickwood, ed.), pp. 77–142. IRL Press at Oxford University Press, Oxford.
17. **Hinton, R.H. and Mullock, B.M.** (1997) in *Subcellular Fractionation – A Practical Approach* (J.M. Graham and D. Rickwood, eds), pp. 31–69. IRL Press at Oxford University Press, Oxford.
18. **Ford, T.C., Graham, J.M. and Rickwood, D.** (1994) *Anal. Biochem.* **220:** 360–366.
19. **Gupta, D. and Tartakoff, A.M.** (1989) *Methods Cell Biol.* **31:** 247–263.
20. **Amar-Costesec, A., Wibo, M., Thinès-Sempoux, D., Beaufay, H. and Berthet, J.** (1974) *J. Cell Biol.* **62:** 717–745.
21. **Courtoy, P.J., Quintart, J. and Baudhuin, P.** (1984) *J. Cell Biol.* **98:** 870–876.
22. **Luzio, J. P. and Richardson, P. J.** (1993) in *Methods in Molecular Biology*, Vol. 19 (J. M. Graham and J. A. Higgins, eds), pp. 141–151. Humana Press, Totowa, NJ.
23. **Sztul, E.S., Howell, K.E. and Palade, G.E.** (1985) *J. Cell Biol.* **100:** 1255–1261.
24. **Hubbard, A.L., Dunn, W.A., Mueller, S.C. and Bartles, J.R.** (1988) in *Cell Free Analysis of Membrane Traffic* (D.J. Morré, K.E. Howell, G.M. C. Cook and W.H. Evans, eds), pp. 115–127. Alan R. Liss Inc., NY.
25. **Menashi, S., Weintraub, H. and Crawford, N.** (1981) *J. Biol. Chem.* **256:** 4095–4101.
26. **Morré, D.J., Morré, D.M. and Heidrich, H.-G.** (1983) *Eur. J. Cell Biol.* **31:** 263–274.
27. **Harms, E., Kern, H. and Schneider, J.A.** (1980) *Proc. Natl Acad. Sci. USA.* **77:** 6139–6143.
28. **Fisher, D. and Sutherland, I.A.** (1989) *Separations Using Aqueous Phase Systems. Applications in Cell Biology and Biotechnology.* Plenum Press, New York.
29. **Graham, J.M.** (1993) in *Methods in Molecular Biology*, Vol. 19 (J. M. Graham and J. A. Higgins, eds), pp. 1–18. Humana Press, Totowa, NJ.

30. Appendix 4 in *Preparative Centrifugation – A Practical Approach* (D. Rickwood, ed.), pp. 369–387. IRL Press at Oxford University Press, Oxford.
31. **Zaar, K., Völkl, A. and Fahimi, D.** (1989) *Biochem. J.* **261:** 233–238.
32. **Draye, J.-P., Courtoy, P.J., Quintart, J. and Baudhuin, P.** (1987) *Eur. J. Biochem.* **170:** 405–411.
33. **Morreé, D.J., Cheetham, R.D. and Nyquist, S.E.** (1972) *Prep. Biochem.* **2:** 61–67.
34. **Lodish, H.F., Kong, N., Hirani, S. and Rasmussen, J.** (1987) *J. Cell Biol.* **104:** 221–230.
35. **Gjøen, T., Berg, T.O. and Berg, T.** (1997) in *Subcellular Fractionation – A Practical Approach* (J.M. Graham and D. Rickwood, eds), pp. 169–203. IRL Press at Oxford University Press, Oxford.
36. **Cartwright, I.J., Higgins, J.A., Wilkinson, J., Bellavia, S., Kendrick, J.S. and Graham, J.M.** (1997) *J. Lipid Res.* **38:** 531–545.
37. **Graham, J.M. and Billington, D.** (1996) *Z. Gastroenterol.* **34** (Suppl. 3): 76–78.

4 Molecular composition

This chapter will deal with the basic analysis of the three principal ingredients of all membranes: protein, lipid and carbohydrate. It will concentrate on the techniques which answer the questions 'how much?' and 'what type?'

4.1 Chemical analysis of proteins

4.1.1 Quantitative assay

A reliable estimation of the protein content of a membrane fraction is essential for quantitation of other components, for example for the expression of an enzyme concentration as a specific activity (e.g. µmol substrate reacted h^{-1} mg^{-1} protein) and it is often necesssary to control the amount of membrane used in a subsequent procedure (e.g. electrophoresis or reaction with antibodies). Although the actual assay may be very simple, there are a number of important factors that may influence the choice of assay and also the accuracy of the data. There are several spectrophotometric assays for proteins whose applicability depends to a large extent on the samples.

Gradient samples which contain an increasing concentration of an interfering solute can be a particular problem. To avoid this it may be necessary to precipitate the protein with trichloroacetic acid or to sediment the membranes by centrifugation after dilution of the gradient fraction with a buffer and then to resuspend the membrane pellet in water.

Many assays require that the membrane protein is solubilized, either in an alkaline solution or in sodium dodecyl sulfate (SDS). The level of SDS used must not cause serious interference with the assay.

Folin–Ciocalteu method. Most protein assays have been designed for the estimation of soluble proteins and the most widely used method, which has been available for more than 40 years, is that based on the use of the Folin–Ciocalteu phenol reagent (Box 4.1). Although this is a common method, the precise reactions involved are not clearly understood. It detects principally tyrosine residues and it is thought that the blue product is a result of the reduction of Cu^{2+} to Cu^+ at the peptide bond site, followed by reaction with the Folin–Ciocalteu reagent [1]. Although the reagent is unstable in the alkaline conditions necessary for the Cu–peptide bond reaction and it needs to be carried out in two steps, it remains a popular assay. It has been adapted to membrane samples and its sensitivity increased [2].

Box 4.1

Folin–Ciocalteu method for proteins

Principles: The standard procedure uses four reagents: (a) 2% (w/v) sodium carbonate in 0.05 M NaOH; (b) 2% (w/v) sodium potassium tartrate; (c) 1% (w/v) copper sulphate; and (d) the Folin–Ciocalteu reagent which is widely available commercially [2].

The NaOH solubilizes the membrane and provides the necessary alkaline conditions for the reaction and the tartrate stabilizes the Cu^{2+} in alkaline solution.

The colored product develops over a period of 45 min and is stable for another 1 h.

Sensitivity: The method is satisfactory for 5–50 µg protein.

Interference: Sucrose and iodinated density gradient media at concentrations above 10% (w/v) and various organic buffers (but only above 0.2 M). Most membrane samples can be diluted by the assay mixture to concentrations of gradient media which will not interfere.

Detergents may cause precipitation; nevertheless there are a number of modifications which employ low levels of either SDS or deoxycholate to solubilize the membrane (e.g. see [3]).

Bicinchoninic acid assay. A modification of the Folin–Ciocalteu method uses a more stable reagent to complex the Cu^+. Bicinchoninic acid (BCA) has a high affinity and specificity for this anion. It is also not affected by raised temperature and so the method can, if required, be made more sensitive by carrying out the assay at 60°C rather than at room temperature. See Box 4.2.

Coomassie blue methods. Rather than rely on a chemical reaction to produce a colored product, Coomassie blue is used in a simple dye-binding method, first developed by Bradford [5]. When the dye binds to a protein its absorbance maximum is shifted from 465 to 595 nm and it is the increase of A_{595} which is measured. The method is very simple and rapid (see Box 4.3).

Box 4.2

Bicinchoninic acid (BCA) assay for proteins [4].

Principles: The copper solution is the same as for the Folin method; the alkaline solution contains 8% $Na_2CO_3 \cdot H_2O$, 1.6% NaOH, 1.6% sodium tartrate adjusted to pH 11.25 with bicarbonate and the color reagent is 4% BCA.

Sensitivity: If the assay is carried out at 60°C it is satisfactory for samples containing between 5 and 50 µg protein.

Interference: The reaction is much less sensitive to interference from gradient media than the Folin–Ciocalteu method.

Box 4.3

Coomassie blue method for proteins

Principles: The protein solution in 0.1 ml is mixed with 5 ml of 0.01% (w/v) Coomassie blue, 4.7% (w/v) ethanol and 8.5% (w/v) phosphoric acid and after 2 min (but before 1 h) the A_{595} measured against a blank containing no protein.

Sensitivity: The method is quite satisfactory with samples containing protein down to 5 µg.

Interference: Generally unaffected by the presence of sucrose or iodinated density gradient media, but is badly affected by detergents above 1% concentration.

A modification of the Coomassie blue method was introduced by Winterbourne [6]. The membrane sample is spotted on to Whatman 3MM paper, allowed to dry and then stained with a Coomassie blue dye solution. The dye solution is then removed and the paper destained (much like an SDS–PAGE gel) to remove the background stain. The filter paper is then dried; the dyed protein eluted and the A_{595} measured. Although there are more steps than in the Bradford assay, it is more sensitive as the unbound dye is removed before measurement, and it is very easy to carry out. After drying the sample on the paper, it can be washed (fixed) in trichloroacetic acid to precipitate the protein on to the paper and to remove all potentially interfering solutes from the sample. Only the staining step is time-dependent, otherwise the procedure can be interrupted at almost any stage.

The amino acid composition of the protein, in particular its tyrosine content, can influence the yield of colored product quite significantly with the Folin–Ciocalteu and bicinchoninic acid assays; Coomassie blue binding assays are rather less affected by the type of protein used as a standard.

4.1.2 Separation of membrane proteins by sodium dodecyl sulfate–polyacrylamide gel electrophoresis (SDS–PAGE)

The most frequently used method to analyze membrane proteins is to solubilize them by dissociation in SDS and to separate them by electrophoresis on polyacrylamide gels. In the earlier methods, the gels were cast as cylindrical rods in glass tubes (one rod gel being required per sample). Today, however, unless the samples are to be analyzed by two-dimensional electrophoresis (see Section 4.1.4), this technique has been superceded by slab gels (contained between two glass plates) which can accommodate multiple samples and thus make comparative analysis of samples much easier. Samples are applied to the gel in a series of wells which are formed by the insertion of a comb into the polymerizing solution (see *Figure 4.1*). Almost without exception, the discontinuous system of Laemmli [7], based on the methods of Ornstein [8] and Davis [9], is used for membrane proteins. The system consists of a resolving gel (pH 8.8), topped by a stacking gel (pH 6.8) which contains the wells. The top and bottom of the gel are in contact with the electrode buffer so that a potential difference can be applied across the gel. Concentration of the proteins (stacking) at the top of the resolving gel is achieved by the pH difference of the two gels and the differential anion content of the electrode buffer (glycinate) and the gels (chloride). This provides sharp bands in the subsequent resolving gel. A thorough discussion of electrophoresis theory and practical details regarding gel preparation, equipmentation or the running of SDS–PAGE can be found in a number of excellent review articles and books (e.g. see [10]). This short section will concentrate more on sample handling (Box 4.4), choice of conditions and a few specimen results.

Solubilization of membrane proteins by SDS precludes any fractionation of these molecules on the basis of their native charge, since the charge on the protein–SDS complexes is overwhelmingly from the SDS itself (approx. 1.4 g of SDS are bound by 1 g of protein). Separations of membrane proteins by SDS–PAGE therefore occur on the basis of molecular size (or, more precisely, effective molecular radius) rather than charge. A graph of log M_r versus distance moved (see *Figure 4.2*) can be plotted by running a mixture of protein

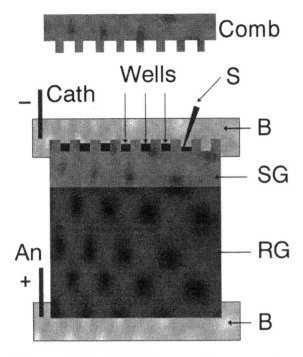

Figure 4.1. SDS–PAGE: diagrammatic representation of slab gel system.
S, Sample applied to wells formed by comb; Cath, cathode; B, electrophoresis
buffer; SG, stacking gel; RG, running gel; An, anode.

standards (of known molecular mass) in one of the wells and this is
used to determine the M_r of any protein band in the sample.

Apart from molecular mass data, SDS–PAGE provides little
information about the identity of the various fractionated polypeptides
or their function. In combination with immunoblotting (Section 4.1.3),
however, it can provide such evidence. The clear allocation of a
function to a specific protein by immunoblotting, however, depends on
the degree of resolution which the fractionation process can achieve.
SDS–PAGE is also used to identify proteins which have been
vectorially labeled with a radiolabeled, fluorescent or chemical probe
(see Chapter 5).

Choice of gel formation conditions will depend on whether a general
survey of membrane proteins is to be carried out or a more precise
study of certain proteins whose M_r is already known. A typical stock
acrylamide solution contains 29.2 g acrylamide and 0.8 g *bis*-
acrylamide in 100 ml water – a so-called 30% T/2.7% C solution, where
T is the total monomer (acrylamide + *bis*-acrylamide) in % (w/v) and C
is the amount of *bis*-acylamide as a % (w/w) of the total monomer

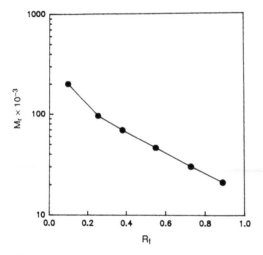

Figure 4.2. Molecular mass calibration of SDS–PAGE. Plot of molecular mass
(M_r) on log scale against distance moved by protein/distance moved by tracker
dye (R_f) on a 10% T SDS–polyacrylamide gel. Reproduced from Dunn, M.J.
(1993) with permission from BIOS Scientific Publishers [10].

Box 4.4

Preparation and application of samples for SDS–PAGE.

Strategy

Sample buffers vary in detailed composition:

1. SDS is invariably present.
2. A reducing agent to break S–S bonds is optional.
3. Glycerol is often included to raise the density of the sample so that it can be
   layered easily under the reservoir buffer. The sucrose, metrizamide, Nycodenz
   or iodixanol
   present in gradient fractions may replace the glycerol. None of these molecules
   are ionic and hence do not interfere with the electrophoresis process. Samples
   containing Percoll, however, are apparently unsuitable for SDS–PAGE.
4. A suitable buffer (normally Tris–HCl).
5. A tracker dye which migrates just ahead of the proteins.
6. To achieve S–S bond reduction and complete solubilization the sample is heated
   (60–100°C).

Typical protocol: Mix membrane suspensions with an equal volume of a buffer
containing 125 mM Tris, pH 6.5, 4% SDS, 0.002% bromophenol blue, 20% glycerol,
10% 2-mercaptoethanol and heat at 100°C for 3 min.

The same amount of protein should be applied to each well of a slab gel, not only to
ensure that the gel is not overloaded but also to make valid comparisons between
different samples.

(acrylamide + *bis*-acrylamide). The final % T required for the optimum separation of proteins in different M_r ranges is given in *Table 4.1*. Some membranes such as that of the human erythrocyte contain relatively few proteins (*Figure 4.3*) while in most other membranes, the protein pattern on SDS–PAGE is much more complex (*Figure 4.4*). The range of protein sizes which can be effectively separated on a single gel is increased by using a gradient gel, and a 5–20% T (%C = 2.6) is able to resolve proteins of M_r = $14–200 \times 10^3$.

Table 4.1. Effect of acrylamide concentration (%T) on resolution in SDS–PAGE (from [10])

% T	Optimum M_r range
3–5	>100 000
5–12	20 000–150 000
10–15	10 000–80 000
>15	<15 000

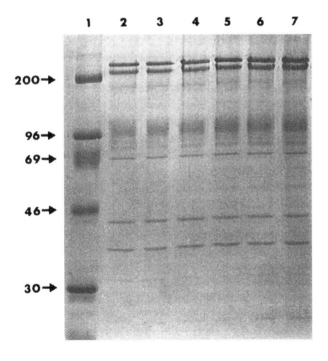

Figure 4.3. Results of 10% T SDS–PAGE of human erythrocyte membrane proteins. Lane 1 contains mol. wt markers and the arrows indicate the M_r values; lanes 2–7, erythrocyte membrane proteins, with 20 µg (2,3), 25 µg (4,5) and 30 µg (6,7) protein loading. Reproduced from Dunn, M.J. and Bradd, S.J. (1993) with permission from Humana Press [11].

Track

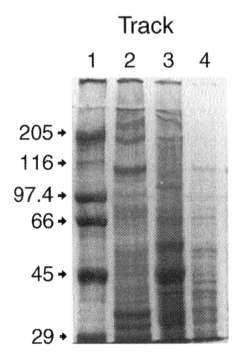

Figure 4.4. 7.5%T SDS–PAGE gel of rat liver membrane fractions. Track 1: molecular weight markers; track 2: plasma membrane (40 μg protein); track 3: Golgi membranes (40 μg protein); track 4: lysosomes (14 μg protein). Gel stained with Coomassie blue.

Glycoproteins. The estimation of M_r of membrane proteins on SDS–PAGE gels by comparison with the electrophoretic mobility of markers of known M_r can be problematical as many of the important membrane proteins are glycosylated. Glycoproteins tend to run anomalously in SDS–PAGE because they bind less SDS per gram than do unglycosylated molecules. As the major contribution to the charge of the SDS–protein complex is the SDS, this will lead to a reduced charge:mass ratio. Glycoproteins thus move rather less quickly than they should do and consequently have a higher apparent M_r.

- The anomalous migration of glycosylated proteins in SDS–PAGE is minimized by using a Tris–borate–EDTA buffer; the borate–carbohydrate complex effectively increasing the overall charge on the molecule [10].

Detection. A variety of chemical stains are available for visualization of the protein bands.

Coomassie blue: The protein load should be 50–100 μg per track for the standard large format gel. Some of the minor bands may be difficult to detect and some of the major ones may be overloaded. If there is a relatively small number of proteins in the membrane (e.g. the human erythrocyte membrane) rather less will be required (20–30 µg).

Silver staining: A protein load of 10–15 µg per track is normally sufficient.

Periodic acid–Schiff (PAS) stain: Glycoproteins are detected on gels by PAS; the relative insensitivity of this detection method has been improved significantly by combining it with a silver staining technique [10]. Alcian blue, dansyl hydrazine and thymol-sulfuric acid are other possible stains for glycosylated molecules.

Radiolabeling: If the sample has been radiolabeled in some way, then equal numbers of counts per track may be a better way of comparison. [^{14}C]leucine or [^{35}S]methionine are commonly used to label proteins in membranes from cultured cells. For phosphoproteins, $^{32}PO_4^{3-}$ is the obvious choice of label, but great care has to be taken in the handling of such material because of the hazardous nature of the isotope; always seek advice from your Safety Officer before using this isotope. Glycoproteins can be labeled with [^{3}H]glucosamine, while surface proteins labeled with ^{125}I using the lactoperoxidase method (or Iodogen™) on the whole cell provide information about the spatial distribution of proteins (see Chapter 5).

γ-Emitters such as ^{125}I and strong β-emitters such as ^{32}P can be detected by simple autoradiography of the gels, while the weaker emitters, ^{14}C, ^{35}S and ^{3}H, are better detected by fluorography after impregnating the gel with a commercial fluorographic reagent.

Quantitation. For quantification of stained bands on a gel or the bands on an autoradiograph, there is a huge range of laser densitometers, radioisotope imagers, etc. linked to a variety of computer driven analyzers and displays. For a review of these, the reader is referred to [10].

4.1.3 Western blotting (immunoblotting)

A technique which has revolutionized the identification of membrane proteins on gels is Western blotting, sometimes called electroblotting or immunoblotting. This has been greatly assisted by the burgeoning availability of antibodies to these proteins. Proteins are transferred from the polyacrylamide gel, by a current applied perpendicularly to the gel, to a matrix (transfer membrane) in contact with the gel (or sometimes simply by contact diffusion).

There are two types of apparatus. In the vertical type the gel and membrane, sandwiched between filter papers and held within a frame, are immersed in a suitable buffer, between platinum electrodes. In the horizontal type the gel/membrane/filter paper layers are sandwiched between two graphite plate electrodes; the filter papers are soaked in transfer buffer for electrical contact. This is also called semi-dry blotting. The relative merits of the two systems are excellently covered in [10].

Box 4.5

Immunoblotting of SDS–PAGE gels [10].

Protocol

Step 1. Soak the gel in a Tris–glycine transfer buffer to remove as much of the SDS as possible. Inclusion of methanol improves this elution but as it also acts as a fixative it retards electrotransfer, particularly of higher M_r proteins.

Step 2. Electrotransfer the proteins to a nitrocellulose membrane.

Step 3. Once the proteins have been transferred to the membrane, they are effectively immobilized and highly accessible to detection reagents, which are usually antibodies or other ligands.

Practical note: Before exposing the transfer membrane to a ligand, all residual protein-binding sites on the matrix must be blocked by reacting with one of a variety of reagents. A solution of dried non-fat milk has become very popular.

Detection. The blot is probed with an unlabeled antibody (specific to the protein of interest) which is subsequently detected with a 'tagged' secondary antibody, usually the anti-IgG for the species in which the probing antibody was raised (*Figure 4.5*).

> The incubation of the blot with the primary antibody is normally carried out in small volumes in plastic bags to avoid using large amounts of the probe.

Some of the most common detection systems are given in *Table 4.2*. Today there is a trend away from the use of radiolabeled species to more sensitive and less hazardous methods.

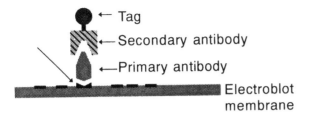

Figure 4.5. Diagrammatic representation of immunoblotting.

Table 4.2. Immunoblotting detection systems

Secondary reagent tag	Mode of detection
^{125}I	Gamma counter
Fluorescein isothiocyanate	Fluorescence
Horseradish peroxidase	Chemical (e.g. H_2O_2 and DAB[a])
Alkaline phosphatase	Chemical (e.g. BCIP[a] and NBT[a])
Biotin	Avidin conjugated to peroxidase or alkaline phosphatase followed by chemical detection as above
Biotin	Streptavidin conjugated as above
Horseradish peroxidase	The O_2^- produced in the presence of peroxide is used to oxidize luminol which emits light when it decays, detected on photographic (X-ray) film. The presence of a phenol can enhance this process (enhanced chemiluminescence)

[a]DAB, diaminobenzidine; BCIP, 5-bromo-4-chloroindoxyl phosphate; NBT, nitroblue tetrazolium.

The enhanced chemiluminescence method (see *Table 4.2*) is more sensitive than any other by at least two orders of magnitude. There are other more complex systems such as triple antibody probing using, for example, peroxidase–antiperoxidase; for more information see [10].

4.1.4 Resolution of proteins by isoelectric focusing and two-dimensional (2D) electrophoresis

In isoelectric focusing (IEF), the proteins are separated in a large pore gel according to their isoelectric point in a pH gradient which is generated by mixing commercially available 'ampholytes' of the appropriate pI value. The latter is commonly carried out in rod gels or on pre-prepared plastic strips, either of which can be subsequently applied to an SDS–PAGE slab gel for 2D electrophoresis [10]. The technique is summarized in *Figure 4.6*.

1

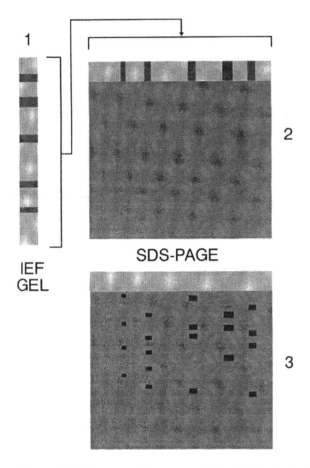

IEF
GEL

SDS-PAGE

2

3

Figure 4.6. Diagrammatic representation of 2D electrophoresis. 1, Proteins separated by IEF (rod gel or strip); 2, rod or strip applied to edge of SDS–PAGE slab gel; 3, proteins separated in second dimension on the basis of size.

- IEF cannot be carried out in the presence of an ionic detergent such as SDS (or at least at the concentrations of this detergent which are routinely used for the solubilization of membrane proteins); nonionic detergents such as Triton X-100, Nonidet P-40 and 3-([3-cholamidopropyl]-dimethylammonio)-1-propanesulfonate (CHAPS) are compatible, as is urea. Consequently the resolving power of IEF and 2D electrophoresis analysis of membrane proteins may be limited.
- Two-dimensional electrophoresis of the hepatocyte plasma membrane resolves 70 or more proteins (*Figure 4.7*).

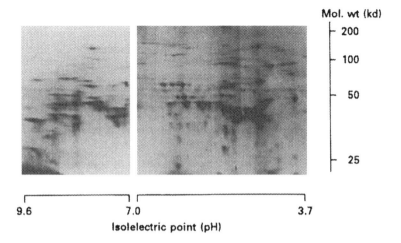

Mol. wt (kd)

— 200

— 100

— 50

— 25

┌──────────────┐ ┌──────────────┐
9.6   7.0   3.7
Isolelectric point (pH)

Figure 4.7. Separation of rat liver plasma membrane proteins by 2D electrophoresis. Reproduced from Evans, W.H. and Graham, J.M. (1989) with permission from IRL Press at Oxford University Press [12].

Box 4.6

Preparation of sample for IEF

A convenient way round the solubility problem is to dissolve the membrane in 1% (w/v) SDS (in water) and then add an equal volume of the IEF sample buffer. A common IEF buffer would be prepared from 6.5 ml water, 5.4 g urea, 0.4 g CHAPS, 0.1 g DTT, plus a 0.5 ml mixture of suitable IEF ampholytes [10].

4.2 Chemical analysis of lipids

The lipids of a membrane (see Chapter 1) have a broad range of polarities: cholesterol is highly apolar, phospholipids are more polar and some glycolipids have extensive water-soluble carbohydrate residues.

4.2.1 Lipid extraction

Membrane lipids are solubilized by one or more organic solvents which also denature the membrane proteins. The most commonly used solvent extraction procedures are those based on Folch *et al.* [13] and a common protocol for membranes is given in Box 4.7.

> **Box 4.7**
>
> **Standard lipid extraction schedule [14].**
>
> **Protocol**
>
> **Step 1.** Stir a membrane suspension in water (< 50 mg dry weight ml^{-1}) with 5 volumes of CHCl$_3$/CH$_3$OH (2:1, v/v) for 5 min.
>
> **Step 2.** Centrifuge to remove the denatured protein. The pellet can be re-extracted to remove more lipid.
>
> **Step 3.** Recentrifuge the combined extracts and evaporate to dryness in a rotary evaporator.
>
> **Step 4.** Redissolve the extract in CHCl$_3$/CH$_3$OH (2:1, v/v); add one-fifth volume of 50 mM CaCl$_2$, mix well and allow the two phases to separate. The lower phase containing lipid is removed and dried down.
>
> **Applicability:** The upper polar phase from step 4 contains nonlipid contaminants and highly polar gangliosides. They can be recovered by dialysis and lyophilization.
>
> For complete extraction of phosphatidylinositol the solvent in step 2 should contain 0.2 ml of conc. HCl per liter (but note that this treatment will hydrolyze glycolipids).
>
> A simplified protocol is to extract the suspension with 20 vol. of solvent and then add one-fifth volume of 50 mM CaCl$_2$ after centrifugation.

> Drying down small volumes of lipids in an organic solvent should always be carried out under a stream of oxygen-free nitrogen; if a rotary evaporator is used then this should be flushed out with nitrogen before use. It is important to prevent possible oxidation of lipids at all stages. Not only is oxidation harmful in itself, it may affect any subsequent separation.

Glycolipids. The extraction of glycolipids follows the same principle as that given in Box 4.7, but the CHCl$_3$/CH$_3$OH (2:1, v/v) extract is mixed with 0.2 vol. 0.1 M KCl and allowed to separate into two phases. Gangliosides partition into the upper layer. The lower layer contains neutral and simple charged glycolipids.

> After the two phases have been separated, it is quite common to wash the lower phase with fresh upper phase solvent and vice versa.

4.2.2 Analysis of phospholipids

Separation. Phospholipids are separated by 1- or 2D thin-layer chromatography (TLC) on glass plates coated with silica gel (Box 4.8).

Although these can be prepared in the laboratory, the process is something of an art and for the occasional user, pre-prepared plates, available commercially from a number of suppliers, are more convenient. They are also available on aluminum or plastic supports.

Box 4.8

Thin-layer chromatography

Strategy: Lipid extracts are dissolved in chloroform/methanol (1:1) and applied in narrow bands to the plate (1). The plate is placed in a tank containing a suitable solvent mixture (S) (2). The solvent is allowed to ascend the plate by capillary action until the solvent front (Sf) is close to the top of the plate (3), thereby separating the lipids according to their polarity.

- In TLC the separation of the phospholipids is achieved by their partition between the stationary polar phase (the shell of water surrounding the silica gel particles) and the mobile solvent (low polarity) phase; i.e. the lower the polarity of the phospholipid, the further it will run on the TLC plate.

The $CHCl_3/CH_3OH$ (1:1, v/v) solvent used to apply the lipids is important as it does not spread on the plate and thus gives 'tight' bands. A 5–10% solution of lipid is commonly used as the sample.

The separation of the various phospholipid species depends on the solvent system that is used. *Table 4.3* lists a few of the more common solvent systems and their uses in 1D TLC and *Figure 4.8* shows diagrammatically the relative migration of the major lipid classes in a typical solvent system.

Table 4.3. Some common solvents for 1D TLC of membrane lipids

Solvent system	Separates
1. Hexane/diethylether/glacial acetic acid (90:10:1)	Major lipid classes
2. Chloroform/methanol/glacial acetic acid/water (60:50:1:4)	Major phospholipids
3. Methanol/chloroform/30% aq. ammonia/water (48:40:5:10)	Phospholipids, particularly PI and its derivatives

A more complete separation of the phospholipid classes and minor lipids can be achieved by 2D TLC. In this system the sample has to be applied as a single spot rather than as a line, so rather less material can be applied and multiple samples cannot be run on a single plate – it is therefore expensive. A typical separation is shown in *Figure 4.9.* For other systems and more experimental detail see [14].

Detection. Once the lipids have been separated on a thin-layer plate, they can be detected by spraying the plate with one of a number of agents listed in *Table 4.4. A number of these stains employ potentially toxic reagents and great care should taken in handling them* [14].

Table 4.4. Some of the more common TLC detection agents

Agent	Chemical reaction	Comment
Iodine vapour	Adds across double bonds	Slow but reversible
1,6-Diphenyl-1,3,5-hexatriene	Fluoresces (UV) when lipid bound	Nondestructive
Ammonium molybdate/conc. H₂SO₄	Standard phosphate stain	Hazardous
Ninhydrin	Amine groups (PE and PS)	Hazardous
Dragendorff stain	Choline reactive (PC and SM)	Very hazardous
1-Naphthol	Stains glycolipids	Hazardous

Quantitative assay of phospholipids. Once the lipids have been separated and detected, quantitative analysis can only be performed on these standard TLC plates after scraping off the area of silica gel

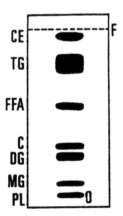

Figure 4.8. Separation of phospholipids by TLC. Solvent system: chloroform/methanol/glacial acetic acid/water (60:50:1:4). NL, Neutral lipid; DPG, diphosphatidylglycerol; PE, phosphatidylethanolamine; PC, phosphatidylcholine; PI, phosphatidylinositol; SM, sphingomyelin; LPL, lysophospholipid; O, origin. Reproduced from Cartwright, I.J. (1993) with permission from Humana Press [14].

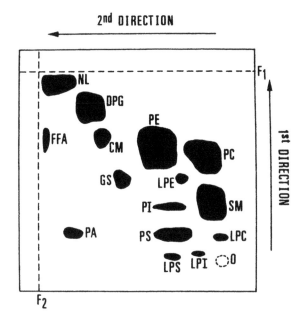

Figure 4.9. Separation of phospholipids by 2D TLC. Solvent system: chloroform/methanol/30% aq. ammonia/water (90:54:5.5:5.5) and chloroform/methanol/acetone/glacial acetic acid/water (60:20:80:20:10). LPC, Lyso-PC; LPE, lyso-PE; LPI, lyso-PI; LPS, lyso-PS; GS, gangliosides; PA, phosphatidic acid; FFA, free fatty acids; CM, ceramide monohexoside. For other abbreviations see *Figure 4.8*. Reproduced from Cartwright, I.J. (1993) with permission from Humana Press [14].

containing each spot (or band) and elution of the lipids into a suitable solvent (Box 4.9). This can only be carried out after use of a nondestructive detection method.

Separation and analysis of phospholipids by HPTLC. The technology of high performance liquid chromatography (HPLC) in which lipids (and other molecules) are separated on columns of very fine grain size adsorbents (see Section 4.2.3) has been applied to the thin-layer method to produce high performance thin-layer chromatography (HPTLC) which provides a much faster and more efficient separation of phospholipids. Commercially available plates using a glass support are recommended to avoid the distortion which can occur with flexible supports.

The system in Box 4.10 demonstrates a practical strategy which is often used to improve resolution without having to use 2D TLC. The first solvent is used to resolve the more polar lipid, and the second solvent resolves further less polar lipids which were poorly resolved in the first solvent (see [14]).

Box 4.9

Quantitation of phospholipids on standard TLC plates

Protocol

Step 1. Locate the lipid spots using a nondestructive method.

Step 2. Scrape off the spots into a glass tube and elute with solvent (usually the same as that used to apply the sample to the TLC plate).

Step 3. Remove the silica gel by centrifugation and dry down the extract.

Step 4. Hydrolyze the lipid in 70% perchloric acid at 230°C, and measure the released inorganic phosphate using a molybdate–Fiske Subbarow method [14].

Practical note: Phosphatidylcholine + sphingomyelin can be determined by assaying the extract for choline.

Box 4.10

Separation of phospholipids by HPTLC

Strategy: Run the plate firstly (to approx. half way up the plate) in chloroform/methanol/glacial acetic acid/ formic acid/water (35:15:6:2:1), and then after drying redevelop in hexane/diisopropyletheher/glacial acetic acid (65:35:2) in the same dimension (to the top of the plate).

HPTLC plates are often used in conjunction with a scanning densitometer after staining the plates with 3% (w/v) cupric acetate in 7% phosphoric acid (with heating to 140°C).

4.2.3 Separation and analysis of phospholipids by HPLC

High performance liquid chromatography (HPLC) was developed to increase the efficiency and sensitivity of column chromatography of a variety of organic compounds, using high pressure solvent pumping and pre-prepared columns containing fine-grain adsorbents. In combination with high resolution automated continuous flow detectors (which use either refractive index or UV absorbance), solvents of high purity and computerized control of solvent profiles, the technique allows high resolution of very small amounts of material. The equipment and consumables are very expensive. Phospholipids can be separated by *normal phase*, i.e. like the TLC systems described in Section 4.2.2, or *reverse phase* in which the column is the nonpolar phase.

• *Normal-phase HPLC* separates phospholipids according to their relative polarities. *Reverse-phase columns* are able to fractionate a single phospholipid class on the basis of fatty acid chain length and unsaturation.

Separations are commonly performed using an 'isocratic' system in which a single solvent is used to elute the phospholipids from the column: the time taken for each phospholipid to be detected at the monitor is highly reproducible and characteristic for each phospholipid. A typical elution profile is shown in *Figure 4.10* and [16] provides details of the equipment and solvent systems.

4.2.4 Analysis of phospholipid fatty acid

Although HPLC can be used to fractionate phospholipids on the basis of their fatty acyl components, gas–liquid chromatography (GLC) is probably the most widely used method for complete fatty acid analysis. Separation of different fatty acids depends upon a partition between a liquid immobilized on the surface of fine particles (the stationary phase) and a gas (the mobile phase). The gas is commonly argon and the stationary phase is a silica powder coated with a wax which becomes a liquid at the operating temperature of the GLC (approx. 250°C). There are a variety of commercially available materials for the stationary phase and columns can be loaded quite easily and cheaply by the operator, although very high resolution pre-prepared capillary columns are now available [17].

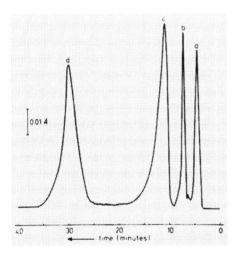

Figure 4.10. Isocratic separation of synthetic phospholipids by HPLC. Solvent: hexane/2-propanol/water. Peak a, PA; b, PE; c, phosphatidylglycerol; d, PC. For abbreviations see *Figures 4.8* and *4.9*. Reprinted from Biochimica Biophysica Acta, 486, Guerts van Kessel, W.S.M., Hax, W.M.A., Demel, R.A. and De Gier, J. 524–530 (1977) with kind permission from Elsevier Science-NL, Sara Burgerhartstraat 25, 1055 KV, Amsterdam, The Netherlands [15].

For GLC, fatty acids have to be converted to a vaporizable form – their methyl ester derivatives. The formation of fatty acid methyl esters (FAMEs) can be carried out on the intact phospholipid molecule (hydrolysis to form the free fatty acids is not necessary); moreover, this transesterification process can be carried out in the presence of silica gel (Box 4.11).

Box 4.11

Strategy for analysis of phospholipid fatty acid profiles [17].

Step 1. Separate the phospholipids by TLC (see Section 4.2.2).

Step 2. Detect the phospholipids using a nondestructive method and scrape the bands off the plate.

Step 3. Methylate the phospholipid fatty acids by heating to 90°C with 12–14% boron trifluoride in methanol, with dichloromethane to improve the solubility of the lipids in the rather polar solvent.

Step 4. Extract the FAME into hexane containing butylated hydroxytoluene (BHT) as an antioxidant.

Step 5. Apply to a suitable GLC system (see [17]).

Important note: It is very important to prevent oxidation of the fatty acids and all the reactions and evaporation of solvents are carried out in an atmosphere of nitrogen.

As the FAMEs emerge from the column they are detected by ionization in a hydrogen/air diffusion flame or by mass spectrometry. As the number of carbon atoms in the fatty acid increases and also as the number of double bonds increases, the FAME partitions more into the liquid (wax) phase and so the retention time on the column increases. The retention times are expressed relative to a standard, often the FAME of stearic acid (18:0). Each peak is then identified by comparison of its relative retention time with those of FAMEs produced from known mixtures of fatty acids. *Figure 4.11* shows the separation of FAMEs from phosphatidylserine in the human erythrocyte membrane.

4.2.5 Determination of cholesterol

Chemical methods for the estimation of cholesterol rely on a reaction involving the production of highly colored products in the presence of strong acids and metal ions (Box 4.12).

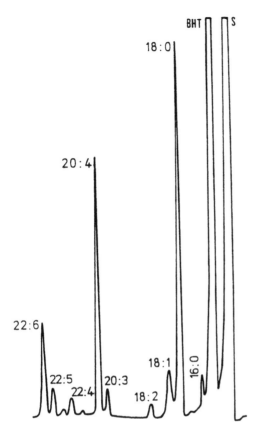

Figure 4.11. Separation of FAME derivatives of fatty acids of human erythrocyte PS by GLC. The elution time increases from right to left; first two peaks are S = solvent; BHT = butylated hydroxytoluene (antioxidant). Fatty acids are eluted by a controlled temperature gradient from195 to 245°C. Reproduced from Cartwright, I.J. (1993) with permission from Humana Press [17].

Box 4.12

Chemical assay of cholesterol

Strategy: With a lipid extract in $CHCl_3/CH_3OH$, the solvent is evaporated and the residue dissolved in iso-propanol or a membrane suspension can be extracted directly with ethanol (1 vol. + 19 vol.). After centrifugation this extract can be used directly in the chemical assay.

In the most commonly used chemical assay the color is developed with ferric chloride in ethyl acetate by the addition of conc. sulfuric acid [18].

Practical note: The addition of the acid is critical and if incorrect may produce charring rather the required colored product.

Enzymic methods are much more satisfactory. Cholesterol oxidase oxidizes the cholesterol to cholest-4-en-3-one. This can be measured directly by its absorbance at 232 nm (Box 4.13) after extraction into iso-octane (2,2,4-trimethylpentane) or coupled to a color-producing system (a number of these are commercially available). There are three major advantages of the enzymic method: it is far more sensitive than the chemical method, it is easier to carry out and it does not require the production of a lipid extract. The membrane is simply solubilized in a suitable detergent [18].

Box 4.13

Enzymic measurement of cholesterol [19]: direct assay of cholest-4-en-3-one

Step 1. Incubate the membrane suspension in a phosphate buffer with cholesterol oxidase and Lutensol at 50°C for 15 min.

Step 2. Shake the mixture with iso-octane and after a short centrifugation (1000g for 10 min) remove the upper layer and measure the A_{232} in a quartz cell.

4.2.5 Separation and analysis of glycolipids

After lipid extraction of the sample (Section 4.2.1) the total glycolipids from each phase are purified using column chromatography.

Lower-phase glycolipids. Before separation from the other nonglycosylated lipids (cholesterol and phospholipids) in the lower phase, the glycolipids must be acetylated (see [20] and Box 4.14).

Box 4.14

Fractionation of neutral and simple charged glycolipids

Strategy

Step 1. Acetylate the glycolipids using acetic anhydride in pyridine.

Step 2. Fractionate the lipids on a small column of florisil by eluting the column with solvents of increasing polarity (dichloroethane mixed with (in increasing polarity) acetone, methanol or methanol + water). Cholesterol is eluted first, followed by the acetylated glycolipids, followed by the phospholipids.

Step 3. Deacetylate the glycolipids with sodium methoxide.

Upper-phase gangliosides. The choice of column and solvent system depends very much on the precise requirements of the operator; for example, a reverse-phase column (see Section 4.2.3) can be used to remove the more polar components from the upper-phase solvent if the lipids of interest are gangliosides and complex neutral lipids, or a DEAE–Sephadex column may be used for isolating the gangliosides and other acidic lipids. Multistep solvent elutions or solvent gradients can fractionate glycolipids according to neuraminic acid content. This complex issue is addressed in [20].

Fractionation of glycolipids. Glycolipid fractionation is carried out by TLC or by HPTLC on glass, plastic or aluminum-backed sheets. The most widely used solvent system is $CHCl_3/CH_3OH/0.5\%$ $CaCl_2$ (60:40:9). For fine resolution of ganglioside mixtures it is sometimes recommended that after running the plate in the solvent, it is allowed to dry and then re-run in fresh solvent in the same direction [20]. A typical separation of gangliosides on HPTLC is shown in *Figure 4.12.*

Detection of glycolipids on TLC plates. A number of chemical sprays are avilable, all of which are potentially corrosive.

- A general glycolipid spray for TLC plates is based on the use of orcinol which reacts with carbohydrate residues; the spray reagent contains 0.2% orcinol in 2 M sulfuric acid, and the color is developed by heating at 120°C.
- A ganglioside-specific spray using resorcinol, cupric sulfate and HCl. Again heat is required for color development.

A relatively new approach is to use antibodies to the carbohydrate residues of glycolipids, the principles of detection being essentially similar to those of the immunoblotting of SDS–PAGE gels (Box 4.15). The use of this technique is potentially very powerful and the amount of glycolipid which can be detected is much lower than that detected by chemical methods (see [20]).

Box 4.15

Use of antibodies to detect glycolipids on TLC plates

Strategy

Step 1. Impregnate the plate with polyisobutylmethacrylate.

Step 2. Block nonspecific sites with bovine serum albumin.

Step 3. Incubate with the antibody.

Step 4. Incubate with secondary antibody, either radiolabeled or covalently linked to gold particles, alkaline phosphatase or horseradish peroxidase (see Section 4.1.3).

Quantitation of glycolipids. The quantitation of animal glycolipids is relatively simple as it is based almost entirely on a sphingosine assay.

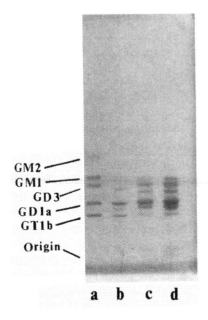

GM2
GM1
GD3
GD1a
GT1b

Origin

a b c d

Figure 4.12. Separation of ganglioside mixtures by HPTLC. Solvent system: chloroform/methanol/0.5% CaCl$_2$ (60:40:9). Detection system: resorcinol spray. For structures of the various gangliosides see Chapter 1. Reproduced from Gregson, N.A. (1993) with permission from Humana Press [20].

Sometimes it is also necessary to measure the N-acetylneuraminic acid content; both assays involve hydrolysis of the glycolipid followed by a simple color reaction.

Sphingosine is liberated by heating in 2 N H$_2$SO$_4$ in dioxane, extracted after alkalinization into ethyl acetate and reacted with methyl orange and H$_2$SO$_4$ (see [20]).

N-Acetylneuraminic acid (sialic acid) is released by hydrolysis of the glycolipid with periodic acid and assayed as a colored product with resorcinol (see [20]).

4.3 Glycoprotein oligosaccharide analysis

The development of cDNA technology means that most studies on the amino acid structure of membrane proteins are now deduced from a knowledge of the nucleotide sequence of the corresponding DNA. Thus the necessity of purifying a membrane protein to homogeneity has been bypassed. This is, however, not the case for the determination of the oligosaccharide structure of glycoproteins which requires: (a) isolation of the glycoprotein of interest; (b) preparation of the oligosaccharide chains; and (c) identification of the sugar sequences. The isolation presupposes that the macromolecule of interest can be detected and identified as a glycosylated protein and this is covered briefly in the next section.

4.3.1 Glycoprotein analysis

Detection and identification of membrane glycoproteins is chiefly carried out by SDS–PAGE in combination with radiolabeling or more commonly with electroblotting and subsequent probing with antibodies coupled to some detection system (see Sections 4.1.2 and 4.1.3). Determination of the molecular mass of membrane glycoproteins by SDS–PAGE is highly problematical (see Section 4.1.2); generally the higher the concentration of acrylamide used, the more reliable is the molecular mass data [10]. Any negatively charged sialic acid may itself contribute to the electrophoretic mobility of a glycoprotein; this can be checked by treating the molecule with neuraminidase, which should cause a reduction in migration. Variation in oligosaccharide structure, including variable sialylation, also leads to the production of rather broad bands in gels.

A simple way of determining whether a particular protein is glycosylated is to prepare an electroblot from an SDS–PAGE gel and then use the strategy outlined in Box 4.16.

Box 4.16

Identification of glycoproteins on electroblots

Strategy

Step 1. Generate free aldehyde groups from the sugars by oxidation with periodate.

Step 2. React the free aldehyde groups with biotin-hydrazide.

Step 3. Probe the blots with an avidin or streptavidin conjugated enzyme and an appropriate chromogen (see Section 4.1.3).

4.3.2 Preparation of a crude glycoprotein fraction

The routine approach is to prepare a crude protein extract from a tissue or whole cells at 4°C (Box 4.17) and then use a targetted selection process (e.g. affinity chromatography using immobilized monoclonal antibodies). Even if the glycoprotein is concentrated in a specific membrane fraction, it is best to avoid lengthy fractionation procedures in case hydrolytic enzymes are released from lysosomes.

Box 4.17

Preparation of a crude glycoprotein fraction [21]

Strategy: Prepare a total protein extract of a cell or tissue by homogenization in an isotonic solution of 2% (w/v) Nonidet P-40 (NP-40). After low-speed centrifugation to remove nuclei and debris, add EDTA and SDS (to 5 mM and 0.4%, respectively) to solubilize the glycoproteins. Remove any insluoble material at 100 000g (1 h).

Or: Produce a crude plasma membrane fraction by homogenization in isotonic 2% (w/v) Tween 20. Prepare a microsomal fraction (see Chapter 3) and then solubilize in 1% (w/v) deoxycholate.

4.3.3 Glycoprotein fractionation

Lectins. Lectins are naturally occurring plant molecules which bind sugars with varying degrees of specificity. Preliminary fractionation of solubilized glycoproteins is frequently carried out by affinity chromatography using lectins immobilized by linking to agarose. *Table 4.5* gives the sugar specificity of some commonly used lectins. Immobilized lectins are widely available commercially, easy to use and good recoveries and concentration of selected glycoproteins are possible [21]. The advent of biotinylated lectins has greatly facilitated their immobilization [22].

Table 4.5. Sugar specificity of some commonly used lectins

Lectin	Sugar specificity[a]
Concanavalin A	α-Man; α-Glc
Lens culinaris	α-Man
Phaseolus limensis	GalNAc
Pokeweed	(GlcNAc)$_3$
Ricin RCA$_{60}$	GalNAc, Gal
Ricin RCA$_{120}$	β-Gal
Soybean	GalNAc
Wheat-germ (WGA)	(GlcNAc)$_2$, Neu5Ac

[a]For abbreviations see Table 1.6.

Because many glycoproteins share some common oligosaccharide sequences, however, this step should be regarded only as a preliminary one. There is also the contrary problem that an individual glycoprotein can exhibit so-called microheterogeneity, i.e. small variations in its oligosaccharide sequences.

It is normal to include a so-called pre-column containing agarose–IgG to reduce nonspecific binding to a lectin column.

The bound glycoproteins are eluted using a high concentration of the sugar specific to the lectin (e.g. 100 mM GlcNAc in the case of WGA) in an appropriate detergent (e.g. deoxycholate).

Deoxycholate is a very commonly used detergent for column chromatography of glycoproteins because it has a high critical micellar concentration (CMC). Detergents with a high CMC are much more easily removed subsequently from the glycoprotein by dialysis than are detergents with a low CMC.

Monoclonal antibodies (mAbs). mAbs covalently bound to Sepharose can also be used for fractionation of glycoproteins. The production of mAbs is very often carried out using the glycoprotein mixture that is purified from the lectin column. For the production, screening and selection of mAbs to a specific glycoprotein see [23].

Gel filtration. Gel filtration is an effective means of separating glycoproteins according to molecular mass. The gel-bead matrix contains pores which are available only to molecules below a certain size; molecules above this size are thus limited to the liquid space between the gel beads. The **exclusion limit** of a gel is defined as the molecular mass of the smallest molecule that is unable to diffuse into the gel matrix pore. All molecules of mass above this limit will therefore have access only to the **void volume** of the gel. Thus when a mixture of macromolecules is applied to the column all molecules of mass greater than the exclusion limit will elute from the column in a single zone, while smaller particles will be retained to an extent inversely proportional to their size and will thus be eluted at a rate approximately equal to $1/M_r$. There is a huge range of gel filtration media and the reader should consult the relevant commercial literature.

Gel filtration is often used as a secondary fractionation to a lectin column. For example a concanavalin A–Sepharose affinity column binds a number of glycoproteins from mouse thymocytes, but the Thy-1 glycoprotein can be effectively resolved from this group of glycosylated proteins on Sephacryl-200™ [21] since its molecular mass is significantly different from the other glycoproteins.

- Gel filtration is also an effective method for substituting buffers or detergents and for removal of small molecular mass contaminants from the purified glycoprotein.

Ion exchange chromatography. This technique can be used for purifying charged glycoproteins (i.e. those containing sialic acid) but highly glycosylated glycoproteins often exhibit variations in the degree of sialylation, which produces broad or multiple bands in column elution profiles.

4.3.4 Chemical assays for carbohydrates

Once the glycoprotein has been purified, the oligosaccharide chains can be totally hydrolyzed either enzymically or by acid hydrolysis. Enzymic digestion with a range of glycosidases is often carried out after a preliminary digestion with Pronase, as smaller glycopeptides may be more susceptible to glycosidases than the intact glycoprotein. Acid hydrolysis is cheap and effective but some procedures employ rather hazardous conditions (e.g. release of amino sugars requires 4 M HCl at 100°C in sealed tubes under nitrogen) and then the acid has to be removed either by ion exchange or by evaporation over alkali.

A range of chemical reactions producing colored products is available for the determination of total sugars, fucose, hexoses, amino sugars and sialic acid (see *Table 4.6*).

Table 4.6. Chemical reactions for sugars

Sugar	Reaction	Reference
Total hexose sugars	Anthrone reagent	24
Galactose	Galactose oxidase	25
Hexosamines	Elson and Morgan	26
N-Acetylneuraminic acid	Resorcinol	27
	Thiobarbituric acid	28

Because many of the reactions for specific types of monosaccharide suffer from interference from other sugar types (the Elson and Morgan reaction is particularly bad in this respect) the total sugar fraction is often passed down an ion-exchange column to separate the neutral sugars from the hexosamines. Individual neutral sugars and hexosamines can be separated by paper or thin-layer chromatography (see [29] for more details).

- Generally chemical detection of monosaccharides is used for amounts greater than 10 nmol. Gas–liquid chromatography (GLC) is normally used for the 1–10 nmol range while at < 1 nmol, high performance anion-exchange chromatography is the best method (see [30]).

4.3.5 Isolation of the oligosaccharide chains

N-linked oligosaccharide chains are released by hydrolysis with peptide N-glycosidase following denaturation of the protein.

O-linked oligosaccharide chains can be released by hydrolysis with O-glycosidase (endo-α-N-acetylgalactosaminidase), but its narrow specificity means that a preliminary series of digestions with a mixture of exoglycosidases (glycosidases from a range of bacterial, plant and animal sources) is necessary for this enzyme to be effective.

Both O-linked and N-linked oligosaccharide chains are released by hydrazinolysis, and conditions can be chosen to release one or the other preferentially or both types together. The method is not oligosaccharide-sensitive and it has also been automated.

As the release of N-linked oligosaccharides by peptide N-glycosidase occurs in a stepwise manner, then the M_r of the protein also decreases stepwise. If the products of hydrolysis are monitored by SDS–PAGE with time, a series of bands of increasing electrophoretic mobility will be observed and the number of new bands indicates the approximate number of oligosaccharide chains (see [31]).

4.3.6 Oligosaccharide sequence analysis

Mass spectrometry and nuclear magnetic resonance spectroscopy can both provide useful information about sugar sequences of the released oligosaccharide chains, but one of the most widely used techniques which does not rely on the availability of expensive equipment employs exoglycosidases, which hydrolyze the links between certain sugars with high specificity. Nine different types of enzyme are in common use:

- sialidase, α-D-galactosidase, β-D-galactosidase, endo-β-galactosidase, N-acetyl-α-D-glucosaminidase, N-acetyl-β-D-glucosaminidase, α-D-mannosidase, β-D-mannosidase and α-D-fucosidase.

In many cases each of these enzymes is available from different sources and will have different C-atom linkage specificities which permit very fine structural analysis. The methodology depends on the validation of the enzyme specificities using artificial substrates, and each enzyme requires its own individual incubation protocols. For a detailed account the reader is referred to [22] and [31]. An overall strategy is given in Box 4.18.

Box 4.18

Oligosaccharide analysis

Strategy

Step 1. Because the material is often available in relatively small amounts, radiolabeling with $NaB[^3H]_4$ is a common preliminary.

Step 2. Separate acidic and neutral oligosaccharides by ion-exchange chromatography using a Dowex-Cl⁻ column. The neutral fraction is unbound and elute any acidic oligosaccharides with pyridine acetate.

Step 3. Separate oligosaccharides by size using gel filtration on BioGel P4.

Step 4. To be able to analyze the size changes in step 5, calibrate the BioGel P4 sizing column using glucose oligomers (from dextran hydrolysis) and standard oligosaccharides.

Step 5. Digest the separated oligosaccharides sequentially with exoglycosidases, analyzing each stage by gel filtration on BioGel P4. Each time a sugar residue has been removed, the BioGel P4 profile will change as the residue becomes smaller.

Practical note: Because many oligosaccharide chains conform to a basic pattern of sugar structure, intelligent 'guesses' can be made as to the use of a likely sequence of exoglycosidases.

Sequence analysis of the glycosaminoglycan (GAG) chains of proteoglycans. This is achieved by a similar strategy, i.e. by the use of polysaccharide lyases to depolymerize the GAGs at specific linkages, followed by characterization of the released oligosaccharides and then secondary treatment with other linkage-specific reagents. Mapping of the released oligosaccharides can be carried out (as with glycoproteins) by gel filtration but either HPLC or gradient PAGE are often used subsequently to provide more detailed information. The isolation of proteoglycans and GAG analysis require procedures which are often tailor-made to an individual macromolecule (see [32] and [33] for more information).

References

1. **Lowry, O.H., Roseborough, N.J., Farr, A.L. and Randall, R.J.** (1951) *J. Biol. Chem.* **193:** 265–275.
2. **Peterson, G.L.** (1983) *Methods Enzymol.* **91:** 95–119.
3. **Markwell, M.A.K., Haas, S.M., Bieber L.L. and Tolbert N.E.** (1978) *Anal. Biochem.* **87:** 206–210.
4. **Smith, P.K., Krohn, R.I., Hermanson, G.T.** *et al.* (1985) *Anal. Biochem.* **150:** 76–85.
5. **Bradford, M.M.** (1976) *Anal. Biochem.* **72:** 248–254.
6. **Winterbourne, D.J.** (1993) in *Methods in Molecular Biology*, Vol. 19 (J.M. Graham and J.A. Higgins, eds), pp. 197–202. Humana Press, Totowa, NJ.
7. **Laemmli, U.K.** (1970) *Nature* **227:** 680–685.
8. **Ornstein, L.** (1964) *Ann. N.Y. Acad. Sci.* **121:** 321–349.
9. **Davis, B.J.** (1964) *Ann. N.Y. Acad. Sci.* **121:** 404–427.
10. **Dunn, M.J.** (1993) *Gel Electrophoresis: Proteins.* BIOS Scientific Publishers, Oxford.
11. **Dunn, M.J. and Bradd S.J.** (1993) in *Methods in Molecular Biology*, Vol. 19 (J.M. Graham and J.A. Higgins, eds), pp. 203–210. Humana Press, Totowa, NJ.
12. **Evans, W.H. and Graham, J.M.** (1989) *Membrane Structure and Function.* IRL Press at Oxford University Press, Oxford.
13. **Folch, J., Lees, M. and Sloane-Stanley, G.H.** (1957) *J. Biol. Chem.* **226:** 497–509.
14. **Cartwright, I.J.** (1993) in *Methods in Molecular Biology*, Vol. 19 (J.M. Graham and J.A. Higgins, eds), pp. 153–167. Humana Press, Totowa, NJ.
15. **Guerts van Kessel, W.S.M., Hax, W.M.A., Demel, R.A. and De Gier, J.** (1977) *Biochim. Biophys. Acta* **486:** 524–530.
16. **Ahmed, H.** (1993) in *Methods in Molecular Biology*, Vol. 19 (J.M. Graham and J.A. Higgins, eds), pp. 169–177. Humana Press, Totowa, NJ.
17. **Cartwright, I.J.** (1993) in *Methods in Molecular Biology*, Vol. 19 (J.M. Graham and J.A. Higgins, eds), pp. 183–195. Humana Press, Totowa, NJ.
18. **Ahmed, H.** (1993) in *Methods in Molecular Biology*, Vol. 19 (J.M. Graham and J.A. Higgins, eds), pp. 179–182. Humana Press, Totowa, NJ.
19. **Trinder, P.** (1981) *Ann. Clin. Biochem.* **18:** 64–70.
20. **Gregson, N.A.** (1993) in *Methods in Molecular Biology*, Vol. 19 (J.M. Graham and J.A. Higgins, eds), pp. 287–301. Humana Press, Totowa, NJ.
21. **Carlsson, S.R.** (1993) in *Glycobiology a Practical Approach* (M. Fukuda and A. Kobata, eds), pp. 1–26. IRL Press at Oxford Univeristy Press, Oxford.
22. *Tools for Glycobiology*, pp. 41–51. Oxford Glycosystems, Oxford.
23. **Lidell, E. and Weeks, I.** (1995) *Antibody Technology*, pp. 25–44. BIOS Scientific Publishers, Oxford.
24. **Roe, J.H.** (1955) *J. Biol. Chem.* **212:** 335–343.
25. **Roth, H., Segal, S. and Bertoli, D.** (1965) *Anal. Biochem.* **10:** 32–52.
26. **Davidson, E.A.** (1966) *Methods Enzymol.* **8:** 52–60.

27. **Svennerholm, L.** (1957) *Biochim. Biophys. Acta* **24:** 604–611.
28. **Warren, L.** (1959) *J. Biol. Chem.* **234:** 1971–1975.
29. **Cook, G.M.W.** (1976) in *Biochemical Analysis of Membranes* (H. Maddy, ed.), pp. 283–351. Chapman and Hall, London.
30. **Manzi, A.E. and Varki, A.** (1993) in *Glycobiology a Practical Approach* (M. Fukuda and A. Kobata, eds), pp. 27–77. IRL Press at Oxford University Press, Oxford.
31. **Corfield, A.P.** (1993) in *Methods in Molecular Biology,* Vol. 19 (J.M. Graham and J.A. Higgins, eds), pp. 269–286. Humana Press, Totowa, NJ.
32. **Lyon, M.** (1993) in *Methods in Molecular Biology*, Vol. 19 (J.M. Graham and J.A. Higgins, eds), pp. 243–251. Humana Press, Totowa, NJ.
33. **Turnbull, J.E.** (1993) in *Methods in Molecular Biology*, Vol. 19 (J.M. Graham and J.A. Higgins, eds), pp. 253–267. Humana Press, Totowa, NJ.

5 Investigating the topology of membrane proteins

In this chapter methods for determining the ways in which membrane proteins are oriented within the membrane bilayer will be discussed. Proteins may be associated with membranes in a number of ways (*Figure 5.1*):

- peripheral proteins which are not integrated into the structure of the membrane;
- proteins with one or more membrane domains;
- proteins anchored to the membrane through covalently bound lipids including phosphatidylinositol, fatty acyl groups and isoprenoid chains.

Before considering the topology of membrane proteins it is necessary to define the two sides of the membrane bilayer. In eukaryotic cells the plasma membrane can be considered as having an extracellular

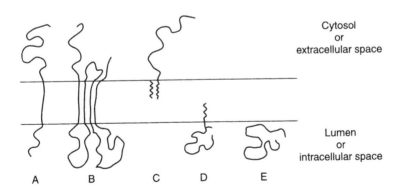

Figure 5.1. Attachment of proteins to membranes. A, Single span proteins – N terminus can be lumenal or in cytosol; B, multispan protein; C, phosphatidylinositol anchor; D, lipid anchor (acyl group or isoprenoid); E, peripheral protein.

surface and a cytosolic surface, while the intracellular organelles have a cytosolic surface and a lumenal or cisternal surface (*Figure 5.2*). When cells or tissues are homogenized the plasma membrane, endoplasmic reticulum and sometimes the Golgi membranes fragment and reseal to form small vesicles (see Chapters 1 and 2). Usually the plasma membrane forms vesicles with the extracellular surface outwards and the endoplasmic reticulum and Golgi membranes form vesicles which have the opposite orientation with the cytosolic surface outwards (*Figure 5.2*). This must be borne in mind when conducting comparative studies of different organelles. The 'self-contained' organelles – mitochondria, lysosomes, nuclei and peroxisomes – all retain the same structure as in the intact cells or tissues.

5.1 Demonstration of the orientation of membrane vesicles

Different methods are available for determining the orientation of membrane vesicles depending on the nature of the vesicle. These include:

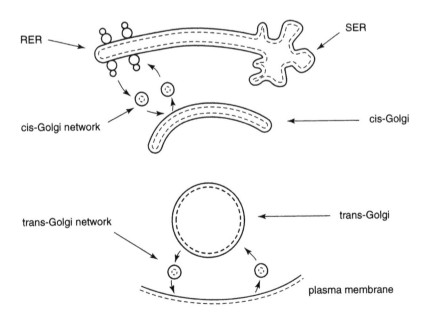

Figure 5.2. Topography of cellular membranes.

- Electron microscopy – Vesicles derived from the RER have bound ribosomes which correspond to the cytosolic side of the membrane *in situ*. Under all homogenization conditions the RER forms vesicles with the cytosolic side outwards. Organelles with recognizable morphology such as Golgi elements containing lipoproteins, mitochondria and lysosomes can be identified using the electron microscope.

- Cytochemical demonstration of enzyme activity at the electron microscopic level – Using either an electron-dense enzyme product or an antibody against a know membrane protein labeled with an electron-dense particle such as gold, the orientation of membrane vesicles can be demonstrated and compared with the same marker used in whole tissue (see Box 5.1).

- Enzyme inhibition – If the membrane vesicles contain enzymes with known orientation, the orientation of the vesicles can be demonstrated using enzyme inhibitors or antibodies against these enzymes (see Box 5.2).

Box 5.1

Aim: To demonstrate the orientation of ER vesicles using glucose-6-phosphatase, which is located at the inner surface of the ER cisternae, as a marker.

Protocol: Glucose-6-phosphatase catalyzes the hydrolysis of glucose-6-phosphate producing glucose and inorganic phosphate. Lead ions capture and precipitate the phosphate at its site of production. To demonstrate glucose-6-phosphatase activity in whole liver, incubate small blocks of liver (fixed by perfusion with 2% glutaraldehyde) with the substrate (glucose-6-phosphate) and the capture reagent (lead nitrate). Conduct control incubations without substrate and/or capture reagent. Wash and refix the tissue blocks in glutaraldehyde followed by osmium tetroxide, embed in plastic and section for electron microscopy. To demonstrate glucose-6-phosphatase activity in isolated microsomal vesicles incubate these in suspension with substrate and capture reagent, pellet by centrifugation, and fix, embed and section the pellet as for the tissue blocks.

Interpretation: Glucose-6-phosphate activity, seen as electron-dense material within the lumen of the ER in whole liver and the lumen of the microsomal vesicles, confirms the orientation of the smooth vesicles.

General applicability: This approach can be used to detect any membrane enzyme, protein or glycoprotein for which a marker is available. If the marker is an antibody this can be labeled with gold or an enzyme such as horseradish peroxidase or biotin–streptavidin (Chapter 3). Glycoproteins can also be detected using similar methods with lectins such as Concanavalin A labeled with an electron-dense marker.

Box 5.2

Aim: 5′-Nucleotidase is a membrane glycoprotein with its active site facing the external medium. Using an antiserum against this protein **it can be demonstrated whether isolated plasma membrane vesicles are of the outside-out or outside-in orientation.**

Protocol: Resuspend the plasma membrane preparation (20–100 μg protein) in Tris–HCl buffer, pH 7.4, and incubate aliquots for 30 min at 30°C with a range of concentrations of anti-5′-nucleotidase serum. Assay the actvity of 5′-nucleotidase by determination of inorganic phosphate released from adenosine monophosphate. Carry out the incubations in the presence and absence of 0.5% deoxycholate to permeabilize the membrane vesicles.

Interpretation: If the vesicles are outside-out then all the 5′-nucleotidase activity will be inhibited by the antiserum in both opened and closed vesicles. If the enzyme is on the inner surface of the vesicles (i.e. they are inside-out) then its activity will only be inhibited in the presence of detergent which opens the membrane.

General applicability: This approach can be used for any enzyme or protein of known orientation using any antibody or other inhibitor that does not penetrate the vesicle. Other examples include galactosyltransferase for the Golgi vesicles and mannose-6-phosphatase for endoplasmic reticulum vesicles. Both of these enzymes are at the inner (lumenal surface) of the membrane and only active in the presence of detergent which opens the vesicles.

5.2 Identification and separation of peripheral and integral proteins

According to the original definition of Singer and Nicolson, membranes contain either **integral** (or intrinsic) **proteins**, which are hydrophobic and anchored through transmembrane domains or by covalent attachment to lipids (phosphatidylinositol, fatty acyl or isoprenyl groups) in the membrane bilayer, or **peripheral** (or extrinsic) **proteins**, which are associated with the membrane through electrostatic interaction usually with integral membrane proteins. The latter group of proteins may be removed without disruption of the membrane structure by relatively mild measures such as treatment with high salt concentrations (>0.15 M), low salt concentrations (<0 mM) or solutions at nonphysiological pH (pH 3–5 or pH 8–12) (see Box 5.3). Integral proteins may only be isolated by disrupting the membrane structure, usually using a detergent.

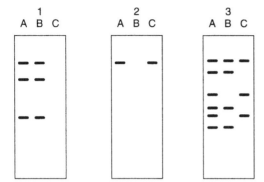

Figure 5.3. Separation of integral and peripheral membrane proteins by phase separation using Triton X114. A, Original membranes; B, aqueous phase; C, detergent phase. 1, Soluble proteins (albumin, catalase, ovalbumin); 2, integral proteins (red cell acetylcholinesterase); 3, RER membranes. The proteins were separated using Triton X114 as described in the text, further separated by SDS–PAGE and stained with Coomassie blue. (Based on [1].)

5.2.1 *Triton X114 as a reagent for separation of integral and peripheral proteins*

The nonionic detergent Triton X114 provides a simple method for separation of the membrane integral and peripheral proteins [1]. The membrane preparation is dissolved in the detergent at low temperature. The integral membrane proteins bind the nonionic detergent, which replaces the membrane lipid, while the hydrophilic peripheral proteins do not interact with the detergent. When the temperature is raised above the cloud point, phase separation occurs. The integral, hydrophobic proteins remain in the detergent-rich lower phase and the peripheral proteins move into the upper aqueous phase. The distribution of the protein of interest can then be determined using SDS–PAGE or immunoblotting. This is not an absolute test as some peripheral proteins behave anomalously; however, taken with the criteria above (removal from the membrane by changes in ionic strength or pH) this is a good indication of whether a protein is peripheral or integral (see Box 5.4).

5.2.2 *Identification of proteins anchored by phosphatidylinositol*

A large number of membrane proteins are anchored to biological membranes through a link with phosphatidylinositol (GPI-) in the

Box 5.3

Aim: To remove and identify the peripheral proteins of red blood cell membranes.

Protocol: Suspend one volume of packed red blood cell ghosts in seven volumes of 0.1 M NaCl or distilled water adjusted to pH 10 or 11. Pellet the membranes by centrifugation at 10 000g for 30 min and analyze the proteins of the supernatant and membrane pellet by SDS–PAGE.

Interpretation: The peripheral proteins are in the supernatant and the integral proteins are in the pellet fractions.

Box 5.4

Aim: To separate integral and peripheral proteins of membrane vesicles.

Protocol

Step 1. Suspend the membrane preparation (2–10 mg ml^{-1}) in Tris–HCl, pH 7.4, 150 mM NaCl containing protease inhibitors (see Chaper 2) and transfer an aliquot (100 µl) into a 1.5 ml Eppendorf tube cooled by immersion in an ice-water bath.

Step 2. Add Triton X114 (0.5 ml of 1%) to give a final concentration of 0.5% and allow the tubes to stand in the ice-bath for about an hour until the membrane dissolves.

Step 3. Centrifuge the tubes ensuring that the temperature remains below 5°C.

Step 4. Prepare a cushion of 300 µl of 6% (w/v) sucrose, 10mM Tris–HCl, pH 7.4, 150 mM NaCl and 0.06% Triton X114 in 1.5 ml Eppendorf tubes. Carefully layer the membrane solution on top of the cushion.

Step 5. Warm the tubes to 30°C for 5 min and centrifuge them for 5 min at 300g to separate the phases. Three layers are seen: a bottom detergent-rich droplet, the sucrose cushion and an upper aqueous phase.

Step 6. If desired each of the phases can be re-extracted – the detergent phase by adding 200 µl of ice-cold buffer to the pellet and the aqueous phase by adding an equal volume of 1% Triton X114 solution – by mixing, cooling and recentrifugation.

Interpretation: The bottom phase is made up of the detergent containing the hydrophobic (integral) proteins and the top phase is the aqueous layer containing hydrophilic proteins which include peripheral proteins and the proteins in the lumen of the vesicles. Using SDS–PAGE and immunoblotting it is possible to determine which specific proteins are integral and which peripheral (*Figure 5.3*).

General applicability: The method can be used for cells in suspension or monolayers or for larger or smaller amounts of membrane suspension by varying the volumes used. The detergent to protein ratio should be about 5:1.

bilayer [2, 3]. Such a link allows a greater mobility of the protein as phospholipids can move in the two-dimensional phase of the membrane an order of magnitude more rapidly than transmembrane protein molecules. In addition, hydrolysis of the link allows the protein, for example an enzyme, to move into the intracellular space. The protein is attached via a core structure in which the inositol of the lipid is linked to glucosamine which is part of a glycan composed largely of mannose residues (*Figure 5.4*). This terminates in ethanolamine phosphate which is linked to the COOH terminus of the protein.

A first indication of a GPI-anchored protein can be obtained from the amino acid/cDNA sequence if this is known. Such proteins have a signal sequence and a C-terminal GPI addition signal peptide which is cleaved as the preassembled GPI precursor is added in the endoplasmic reticulum [4]. The GPI attachment site is not a consensus sequence but may include the residues Ser, Asp, Cys and Gly. The +1 and +2 residues are cleaved. The cleavage site is followed by a polar domain of 7–12 residues and a row of 10–20 hydrophobic amino acids [4].

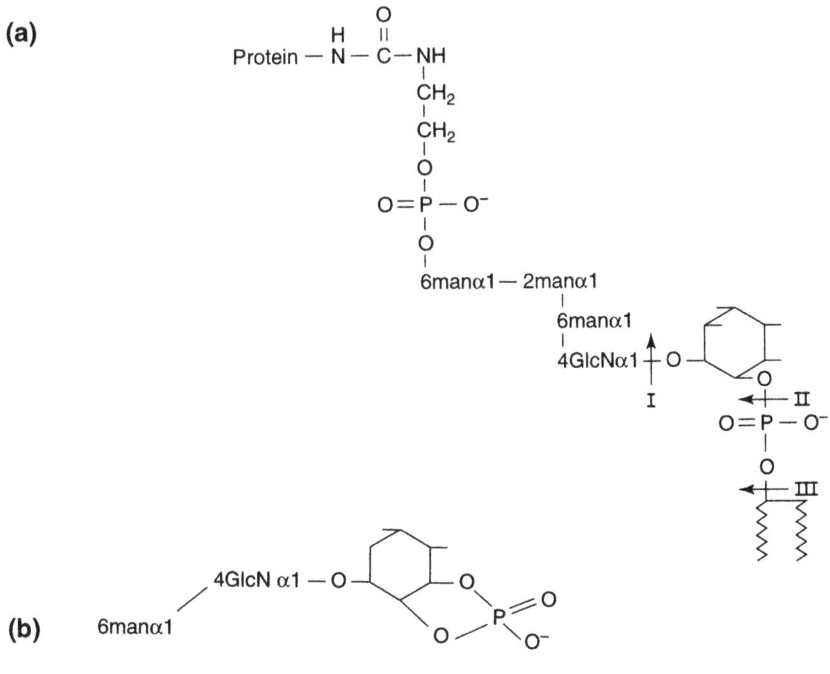

Figure 5.4. (a) Structure of the GPI anchor for membrane proteins. I, Nitrous oxide cleavage site; II, PI-phospholipase C (PLC) cleavage site; III, GPI-phospholipase D cleavage site. (b) CRD epitope exposed by PI-PLC.

After addition of the GPI anchor the proteins are transferred to the plasma membrane and are always oriented towards the extracellular side of the plasma membrane and in polarized epithelial cells have an apical location [5].

GPI proteins can be released from intact cells or membrane vesicles either *chemically*, using nitrous acid or aqueous hydrogen fluoride which cleave the inositol glucosamine bond, or *enzymically*, using a phospholipase C specific for phosphatidylinositol which cleaves between the diglycerol and inositol. Either method can be used in principle; however, the enzymic method is more specific and less likely to damage the membrane or protein. Phosphatidylinositol-specific phospholipase C types have been isolated from bacteria (*Bacillus cereus, B. thuringiensis, Clostridium novyi* and *Staphylococcus aureus*) and from mammalian tissues. The bacterial enzymes have been used most extensively and are now available commercially. To identify GPI proteins the cells or membrane vesicles are treated with the enzyme and protein released detected by pelleting the cells or membrane vesicles and analysis of the supernatant by SDS–PAGE. If antibodies are available specific proteins of interest can be identified by immunoblotting. A scheme for identification of peripheral, integral and GPI apical membrane proteins of cultured cells combining the use of Triton X114 and phospholipase C is shown in Box 5.5. This can be modified for application to tissues and subcellular fractions.

GPI-linked proteins can be unequivocally identified by separation of the membrane or protein under investigation by SDS–PAGE followed by immunoblotting using an antibody to a specific region of the anchor called the cross-reacting determinant (CDR). Such antibodies are available from Oxford Glycosystems UK (see Box 5.6).

5.2.3 Identification of lipid-modified proteins

Membrane proteins may also be modified by covalent attachment of the fatty acids myristate (C14:0) or palmitate (C16:0) or the isoprenoid chains farnesyl (C15) or geranyl-geranyl (C20) [6–8]. Such modification plays a role in the association of the protein with specific membranes and has been implicated in a variety of cellular processes, particularly those involving signaling and regulation. For example, the α subunit of many G-proteins is myristoylated and the γ subunit is prenylated, while the *ras* oncogene protein is prenylated and in some cases palmitoylated.

Myristate is attached to N-terminal glycine co-translationally in the cytosol. The N-terminal methyl group is removed and replaced by myristate, which is donated from its CoA derivative. Myristate remains associated during the life of the protein.

Box 5.5

Aim: To identify GPI proteins at apical surface of monolayers of cultured cells [5].

Protocol

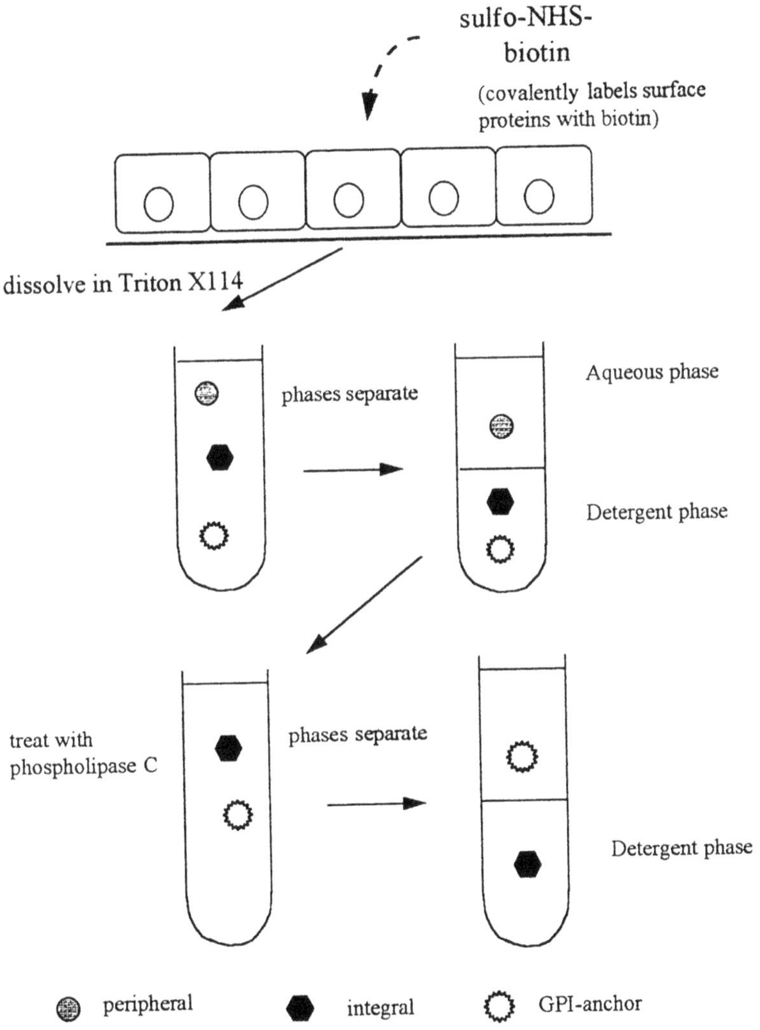

sulfo-NHS-biotin

(covalently labels surface proteins with biotin)

dissolve in Triton X114

phases separate

Aqueous phase

Detergent phase

treat with phospholipase C

phases separate

Detergent phase

peripheral   integral   GPI-anchor

The proteins in each phase are separated by SDS–PAGE and the biotin-labeled bands identified using ^{125}I-streptavidin or streptavidin linked to alkaline phosphatase (see Chapter 3).

General application: This method may be used to detect phosphatidylinositol-linked proteins in cells or organelles in suspension. The biotin may be detected using an enzyme-linked streptavidin and quantitative immunoblotting (see Chapter 4).

Box 5.6

Aim: To identify GPI proteins using anti-CRD.

Protocol: Separate proteins by SDS–PAGE using five lanes plus markers; electrotransfer on to nitrocellulose and cut membrane into lanes.

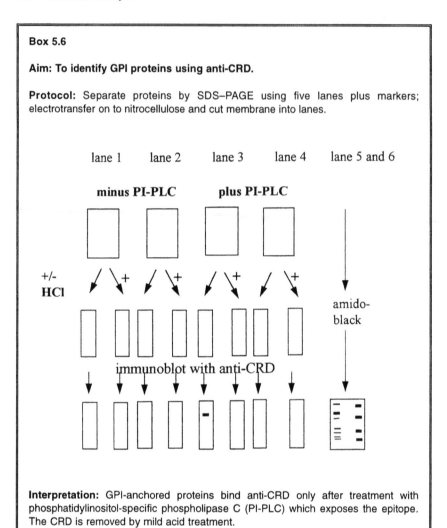

Interpretation: GPI-anchored proteins bind anti-CRD only after treatment with phosphatidylinositol-specific phospholipase C (PI-PLC) which exposes the epitope. The CRD is removed by mild acid treatment.

Palmitate is attached to cysteine post-translationally within the ER/Golgi lumen. Palmityl CoA is the donor. Palmitoylated proteins undergo acylation/deacylation cycles and this probably plays an important role in the function of the protein.

Prenylation involves formation of a thioether link between farnesyl or geranyl-geranyl isoprenoids and cysteine in the sequence CAAX (where A is an aliphatic residue) at the C terminus. After prenylation the terminal amino acids AAX are cleaved and the COOH of the C is methylated. There is evidence that if X is Ala, Met or Ser farnesylation takes place and if X is Leu or Phe then geranyl-geranylation takes place. There are alternative sequences which are also prenylated. C-terminal CC or CCXC are geranyl-geranylated frequently at both terminal cysteines.

Lipid modification of proteins is best detected by metabolic labeling of cells (or bacteria) with radiolabeled lipid followed by analysis of the proteins [4] (see Box 5.7 and Box 5.8).

Box 5.7

Aim: Identification of acylated proteins.

Protocol

Step 1. Label the proteins by incubation of the tissue source (cells or bacteria) with ³H-palmitate or myristate.

Step 2. Dissolve cells in Triton X114 and carry out phase separation as described in Box 5.5. The lipid-modified proteins should move into the detergent phase.

Step 3. Separate proteins by SDS–PAGE and locate the radiolabeled bands *in situ* by counting or fluorography or by cutting out the band, dissolving the gel and counting. Alternatively, acylated proteins may be purified by reverse-phase HPLC using propan-1-ol/water mixtures.

Step 4. The nature of the protein/radiolabeled lipid link can be confirmed either by incubation of protein-containing slices from SDS–PAGE gels or by treatment of proteins in solution as follows:

(i) 0.2 M KOH in methanol for 1 h at room temperature cleaves thio- or oxyesters (palmitate or myristate);
(ii) 1 M hydroxylamine hydrochloride adjusted to pH 7.5 with NaOH cleaves thioesters (palmitate);
(iii) 0.5 M sodium methoxide at 100°C cleaves oxyesters (myristate).

Loss of the radiolabel is detected and quantified by counting the treated protein or by fluorography followed by densitometry.

Step 5. Confirm the nature of the fatty acids released by extraction into chloroform/methanol and analysis by thin-layer chromatography (on Merck plates, with authentic fatty acid standards) or by gas-liquid chromatography (see Chaper 4).

Box 5.8

Aim: Identification of prenylated proteins.

Protocol

Step 1. Label the proteins by incubation of the tissue source (cells or bacteria) with ³H-mevalonate.

Step 2. Dissolve cells in Triton X114 and carry out phase separation as described in Box 5.5. The prenylated proteins should move into the detergent phase.

Step 3. Separate proteins by SDS–PAGE and locate the radiolabeled bands *in situ* by counting, autoradiography or fluorography.

Step 4. Cut out the labeled bands, wash with distilled water to remove SDS and dissolve the protein with pronase (10 µg in 0.1 M ammonium bicarbonate).

Step 5. Extract the radiolabeled isoprenoids and identify these by HPLC.

5.3 Probing the topography of integral membrane proteins

The polypeptide chains of integral membrane proteins loop through the membrane bilayer. The number of transmembrane domains varies from 1 (e.g. glycophorin) to 14 (e.g. band 3 of red blood cells). There are groups of proteins which exhibit extensive homology and have a common orientation, e.g. the G-protein linked receptor proteins which have seven transmembrane domains. For most proteins the transmembrane domain consists of an α-helical sequence of 20–25 amino acid residues; however, the structure of bacterial porin has been solved by X-ray crystallography and consists of a β-barrel made up of six antiparallel strands folded into a transmembrane cylinder lining an aqueous pore [9]. Membrane proteins are difficult to purify and because of their hydrophobic nature are not easily crystallized. Therefore most experimental studies of the orientation of the polypeptide chain with respect to the bilayer have used probes

5.3.1 *General experimental criteria*

The topography of proteins may be probed in subcellular fractions, including membrane vesicles and intact organelles, or in the plasma membranes of intact cells. If cells are under investigation they may be freshly isolated, in suspension culture, or in monolayer culture. In order for results from such investigations to be valid certain criteria must be taken into account in the experimental design. These relate to the specimen under investigation and the probe used.

The membrane or cells under investigation

*   Membrane vesicles or intact organelles must have the same topological orientation (see above).
*   If intact cells are under investigation these must be of the same type and in suspension or in monolayers so that the apical or basolateral surfaces can be probed separately.
*   The vesicles must be closed and impermeable to the probe used.
*   The vesicles, organelles or cells must remain impermeable during the experimental procedures.

The probe

*   The probe must not penetrate the cells or vesicles.
*   The probe should react with known components of the membrane in a detectable and quantifiable way.

5.3.2 Probes for integral membrane proteins

A variety of probes have been developed; these include reagents with low specificity which will react with most proteins:

* Proteolytic enzymes with a range of specificities.
* Chemical labeling reagents usually tagged with a radiolabel to help subsequent detection and reagents with a high degree of specificity to individual proteins.
* Antibodies – polyclonal and monoclonal.
* Enzyme inhibitors.

The choice of probe will depend on the experiment in hand, whether a specific protein is under investigation and whether specific probes such as antibodies are available.

Proteases as probes. A range of proteolytic enzymes with different specificities are available (*Table 5.1*). Conditions of incubation and inhibition of proteolytic activity are provided with the enzyme preparation. In general the most pure preparations available should be purchased, as impurities might cause membrane vesicles to become leaky or cells to become permeable.

The membrane/cell preparation is incubated with and without the protease. At the end of the incubation the enzyme should be inhibited as some proteases will continue to act even in SDS-containing

Table 5.1. Examples of proteases used for probing the orientation of membrane proteins[a]

Enzyme	Type	Inhibitors
Chymotrypsin	Serine endopeptidase	Aprotinin, PMSF, chymostatin, trypsin inhibitor
Elastase	Serine endopeptidase	PMSF, elastinal, α-macroglobulin
Proteinase K	Serine endopeptidase	PMSF, DFP
Trypsin	Serine endopeptidase	Leupeptin, trypsin inhibitor, PMSF, aprotinin, DFP
Endoproteinase Arg-C	Serine protease	DFP, TLCK, α-macroglobulin
Endoproteinase Glu-C	Serine protease	DFP, TLCK, α-macroglobulin
Endoproteinase Lys-C	Serine protease	DFP, aprotinin, leupeptin, TLCK
Papain	Cysteine endopeptidase	SH blocking agents, iodoacetic acid, antipain, aprotinin, PMSF
Pepsin	Aspartate endopeptidase	Pepstatin
Subtilism	Broad specificity	DFP, PMSF, α-macroglobulin
Pronase E	Broad specificity	PMSF
Thermolysin	Zn metalloendopeptidase	EDTA, *o*-phenanthroline, phosphoramidon

[a] This list is not exhaustive: information can be obtained from manufacturers' catalogues (e.g. Boehringer, Sigma, Calbiochem).

solutions. Alternatively, the membrane proteins can be precipitated using trichloroacetic acid which prevents further action of the protease. The membranes are then solubilized, separated by SDS–PAGE and the protein fragments identified by immunoblotting. By comparing gels of membranes incubated with and without different proteases and analysis/sequencing of the fragments, it is possible to determine which proteins are accessible for hydrolysis (see Box 5.9). Control experiments should be performed to demonstrate that proteins not available for hydrolysis in impermeable membrane vesicles or cells become available when the membranes are made permeable or dissolved in detergents.

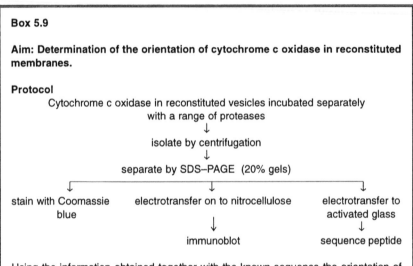

Box 5.9

Aim: Determination of the orientation of cytochrome c oxidase in reconstituted membranes.

Protocol

Cytochrome c oxidase in reconstituted vesicles incubated separately
with a range of proteases
↓
isolate by centrifugation
↓
separate by SDS–PAGE (20% gels)

stain with Coomassie blue	electrotransfer on to nitrocellulose	electrotransfer to activated glass
	↓	↓
	immunoblot	sequence peptide

Using the information obtained together with the known sequence the orientation of the protein in the membrane vesicle can be deduced.

General application: This approach can be used for a number of proteins in different membranes. If monoclonal antibodies are available, these help in the mapping of the peptide fragments. The part of the protein on the inner surface of the membrane vesicle can be probed in opened vesicles, or inside-out vesicles if these are available.

Control experiments must be performed to demonstrate that the membrane vesicles or cells do not become permeable during the incubation. These will vary with the experiment. Suitable controls include:

- For whole cells – Determine the percentage loss of cytosolic enzymes or proteins, e.g. lactic dehydrogenase (for a variety of cells) or hemoglobin (for red blood cells). For example, pellet the cells by centrifugation before and after incubation and assay the lactic dehydrogenase in the supernantant. To determine total (100%) lactic dehydrogenase open the cells with detergent.

- For endoplasmic reticulum or Golgi vesicles – Determine the percentage loss of secretory proteins which are radiolabeled by injection of ^{3}H leucine (100 μC per 100 g body weight) into the portal vein 30 min prior to removal of the liver and preparation of the subcellular fractions. Pellet the vesicles by centrifugation before and after incubation, precipitate the protein with trichloroacetic acid and count the ^{3}H-protein in the pellet. To determine total ^{3}H-protein solubilize the vesicles with detergent, precipitate the protein with trichloroacetic acid and count the ^{3}H-protein.
- For plasma membrane vesicles – Loss of cytosolic enzymes, e.g. lactic dehydrogenase, which are trapped during homogenization.
- For mitochondria or lysosomes – Loss of matrix enzymes.
- For most membrane vesicles – Retention of latency of enzymes known to be located at the lumenal surface (Box 5.10).

Chemical labeling reagents. A variety of reagents are available that react with protein side chains. These include reagents that react with amino acyl, NH_2 groups, COOH groups, SH groups and the OH in tyrosine. Generally, such reagents are most useful if they have a radioactive label to facilitate detection.

The most frequently used methods include:

- Those which make use of two reagents; one which penetrates the membrane vesicle and labels both sides of the membrane and a second that remains outside and labels only the outer surface (see Box 5.11).
- Those using lactoperoxidase, which couples ^{125}I to tyrosine residues in proteins. This procedure is carried out under physiological conditions which helps to preserve the integrity of membrane vesicles or cells. The reaction requires H_2O_2 which is generated by including glucose oxidase in the incubation medium. For studies of membrane protein topography, solid-phase radioiodination using lactoperoxidase coupled to beads (e.g. latex) has the advantage that the coupled enzyme cannot easily penetrate cells or membrane vesicles. Solid-phase lactoperoxidase can be prepared in the laboratory [10] or can be purchased from a number of commercial sources. In the latter case the beads can also be purchased coupled with glucose oxidase in addition to lactoperoxidase (see Box 5.12).

Box 5.10

Aim: Assessment of the integrity of endoplasmic reticulum vesicles by determination of the latency of mannose-6-phosphatase.

Principle: Glucose-6-phosphatase consists of two components, a transporter which carries the substrate across the ER membrane and a hydrolase which hydrolyzes the substrate. The transporter has high specificity and will not transport mannnose-6-phosphate. The hydrolase has lower specificity and will hydrolyze mannose-6-phosphate. Therefore mannose-6-phosphatase is only expressed in leaky vesicles.

Protocol

Step 1. Incubate ER (microsomes) vesicles with mannose-6-phosphate (1 mM) in Tris–HCl buffer, pH 7.4, with and without addition of 0.4% taurocholate for a range of times from 0 to 30 min.

Step 2. Stop the reaction by addition of trichloroacetic acid (final concentration 1%), and pellet the denatured protein by centrifugation in a bench-top centrifuge.

Step 3. Remove aliquots from the supernatant and assay inorganic phosphate released.

Step 4. Ensure that the reaction is linear for the times used. The rate of release of phosphate in the presence of taurocholate (i.e. in opened vesicles) is 100% of the activity. The rate of release of phosphate in the absence of taurocholate as a percentage of that in the presence of taurocholate is a measure of the latency of the enzyme activity.

If microsomal vesicles are closed, less than 10% of the total activity should be expressed in the absence of taurocholate, i.e. the enzyme should be >90% latent.

a. Closed vesicles

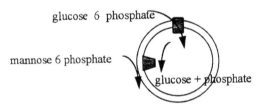

glucose 6 phosphate

mannose 6 phosphate

glucose + phosphate

b. Opened vesicles

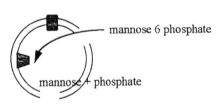

mannose 6 phosphate

mannose + phosphate

 transporter   hydrolase

General applicability: The same principle can be applied to demonstrate the latency of any enzyme which has a lumenal location, e.g. UDP-galactose galactosyltransferase has been used for demonstrating the integrity of Golgi vesicles.

Box 5.11

Aim: Determination of the topography of erythrocyte membrane proteins using radiolabeled probes that react with free amino groups.

Principle: Ethylacetimidate (EA) reacts with free amino groups in proteins and does not penetrate membranes. Isoethionylacetimidate (IEA) reacts with free amino groups and does penetrate the membrane. These will therefore differentially react with extracellular and intracellular protein domains.

Protocol

Step 1. Incubate erythrocytes with ^{14}C-EA or ^{14}C-IEA.

Step 2. Dissolve membranes in sample buffer and separate proteins by SDS–PAGE.

Step 3. Locate radiolabeled protein bands by autoradiography or image analysis.

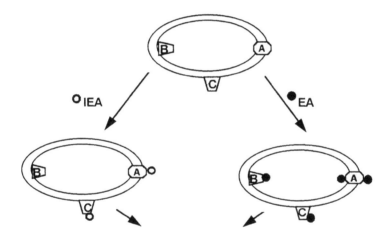

SDS-PAGE AND AUTORADIOGRAPHY

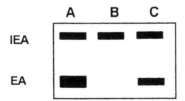

Band A is labeled by both reagents and to a greater extent by EA and is therefore a transmembrane protein with domains on both sides of the membrane.
Band B is labeled only by IEA and therefore lacks an extracellular domain.
Band C is labeled equally by both reagents and therefore lacks an intracellular domain.

General applicability: The same principle can be used to label proteins in any closed vesicles.

Box 5.12

General method for lactoperoxidase (LPO) iodination of cells or membranes.

Protocol

Step 1. Incubate subcellular fractions or cells with LPO, glucose, glucose oxidase and Na^{125}I lactoperoxidase in phosphate-buffered saline, pH 7.4 [10].

Step 2. Remove reagents. Pellet subcellular fractions or cells by centrifugation. Wash several times with buffer containing sodium iodate (0.1 M).

Step 3. Solubilize the membranes or cells in sample buffer, separate by SDS–PAGE and detect the radiolabeled proteins by autoradiography, image analysis or by cutting the band from the gel and counting in a scintillation counter. Alternatively, dissolve the cells or membranes in a nonionic detergent and concentrate and purify the protein of interest by immunoprecipitation before separation by SDS–PAGE.

General applicability: Using similar methods surface-exposed glycoproteins can be radiolabeled by chemical oxidation of the vicinal hydroxide groups in the sugar residues followed by reduction with tritiated borohydride, or all cell or membrane proteins can be labeled with sulf-N-hydroxysuccinimido-biotin (S-NHS-biotin). In the latter case the proteins are separated by SDS–PAGE, electroblotted on to nitrocellulose and the biotinylated proteins are detected using streptavidin linked to alkaline phosphatase as for immunoblots (see Box 5.5) [10].

In these experiments it is essential that the radiolabeling reagents do not gain access to the center of the membrane or the cytosol of cells. Control experiments must be performed to demonstrate that this does not occur. This is best achieved by determining whether a protein known to be in the lumen of the membrane vesicle or the cytosol of the cells becomes labeled. Suitable proteins include:

- for whole cells – lactic dehydrogenase (for a variety of cells) or hemoglobin (for red blood cells);
- for endoplasmic reticulum or Golgi vesicles – secretory proteins, e.g. albumin;
- for plasma membrane vesicles – cytosolic enzymes, e.g. lactic dehyrogenase, which are trapped during homogenization;
- for mitochondria or lysosomes – matrix enzymes.

Chemical labeling of proteins of plasma membrane domains in cultured cells. Cells cultured on monolayers have apical and basolateral surfaces. As this polarity is lost by removal of the cells from the culture dishes, the labeling method (Box 5.11 and 5.12) will identify all the externally exposed proteins. It is also possible to label the proteins at the apical surface specifically or the basolateral surface. The apical surface is exposed to the culture medium and is simply labeled by adding the reagents to the medium followed by detachment of the cells for analysis. To label the basolateral surface, inverted monolayers can be prepared by placing poly-L-lysine-coated coverslips on top of the cultured monolayer and carefully removing them. The apical surface of the cells is bound to the coverslip and thus protected from labeling, while the basolateral surface is accessible to labeling reagents (*Figure 5.5*).

Antibodies as probes. If antibodies are available for the protein under investigation these will greatly facilitate the identification of the protein bands using immunoblotting, which can be used in conjunction with either a protease or radiolabeling to confirm the identity of hydrolyzed proteins, their products or radiolabeled proteins. In addition, antibodies can be used as probes of the orientation of membrane proteins using a technique based on enzyme-linked immunoassay (ELISA). Two general methods are available depending on whether the protein under investigation is available in a purified form (see Box 5.13 and Box 5.14).

ELISA can be used to probe the exposure of a membrane-bound protein at the surface of membrane vesicles (see Box 5.14).

Labeling of the intramembrane part of a protein. The transmembrane domains of a protein can be labeled by allowing a radiolabeled photoactivatable lipid-soluble molecule to partition into the membrane in the dark followed by analysis of the protein [12] (see Box 5.15).

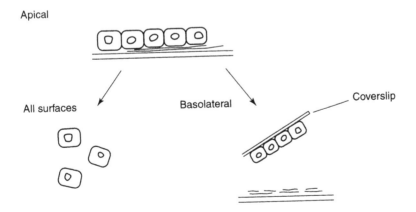

Figure 5.5. Methods for distinguishing apical and basal membrane surface proteins.

Box 5.13

Aim: Probing the topography of proteins by ELISA using a panel of monoclonal antibodies.

Protocol

Step 1. Immobilize membrane vesicles in wells of microtiter plates. Incubate with monoclonal antibodies (mouse IgG). Wash to remove excess antibody.

Step 2. Incubate with secondary antibody (anti-mouse IgG, coupled to alkaline phosphatase). Washy to remove excess antibody.

Step 3. Add alkaline phosphatase substrate (*p*-nitrophenylphosphate) and measure color produced at 550 nm in a plate reader.

Step 4. Repeat assay in the presence of detergent (e.g. 0.5% taurocholate) to open membrane.

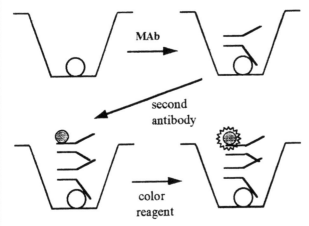

Interpretation: Color development in steps 1–3 indicates that the epitope is exposed. Color development in step 4 indicates that the epitope is intracellular.

Overall comment: Combinations of chemical labeling, proteolytic digestion and immunological probes with sequence analysis provide an overall picture of the way in which a protein is folded into the membrane structure. Examples of proteins solved in this way include bacteriorhodopsin, rhodopsin and Na$^+$/K$^+$-ATPase [13]. A generalized model is illustrated in *Figure 5.6*.

Box 5.14

Aim: Investigation of the topography of apolipoprotein B (apo-B) in subcellular fractions [11].

Principle: This assay is based on competition between immobilized apo-B and unbound apo-B for a limiting amount of monoclonal antibody (MAb; in this example rat IgG). If the epitope recognized is exposed at the outer surface of the membrane vesicles then it will compete with the immobilized protein for the antibody. The bound monoclonal antibody is then quantified using a second antibody coupled to alkaline phosphatase (see Box 5.13 for principle of ELISA).

Protocol

Step 1. Immobilize apo-B (~ 0.5 µg protein) in the form of low-density lipoprotein (LDL) in wells of microtiter plates.

Step 2. Determine the limiting amount of each MAb by performing an antibody dilution curve.

Step 3. Add the selected amount of MAb to each well together with either the subcellular fraction under investigation or a range of concentrations of LDL (to give a standard curve).

Step 4. After incubation, wash the wells and add the second antibody (anti-rat IgG) coupled to alkaline phosphatase.

Step 5. After incubation, wash the wells and measure the amount of second antibody bound by addition of substrate. Read the color development at 550 nm in a plate reader.

Step 6. From the standard competition curve of the log of the competing LDL plotted against the absorbance, calculate the amount of apo-B present in each subcellular fraction.

Step 7. Repeat the assay in the presence of detergent (0.4% taurocholate) to open the vesicles.

Interpretation: It is essential that several concentrations of subcellular fraction are used to ensure that all values obtained lie on the linear part of the competition curve. If the epitope investigated is exposed on the outside of the vesicle then competition will take place. If the epitope is at the inner surface of the vesicle then competition will only occur if the vesicles are opened. If competition occurring with open vesicles exceeds that for closed vesicles this indicates that the epitope may have two locations.

General applicability: This approach can be used for any protein that can be prepared in a relatively pure form for which MAbs are available. If the protein under investigation is not available in a pure form but its topography has been estalished in other membrane vesicles, these can be used as the competitor for such probing experiments. For example, as the topography of the glucose transporter protein in red cell membrane vesicles has been determined, these vesicles can be immobilized to act as competing antigen to investigate whether the topography of the glucose tranporter is similar in other membrane preparations.

Box 5.15

Aim: Intramembranous labeling of red cell glycophorin.

Protocol

Step 1. Add 5-[^{125}I]-iodonaphthyl-1-azide (INA) (20 μl of 1.3 Ci/mmol in ethanol; final concentration 1 μM) to 2.0 ml of red cell ghosts (2 mg protein) and incubate the mixture in the dark for 30 min at 37°C.

Step 2. Irradiate the mixture at 314 nm for 6 min to activate the labeling reagent.

Step 3. Add sample buffer to an aliquot of the reaction mixture, separate by SDS–PAGE and identify the radiolabeled protein by autoradiography or image analysis to check labeling.

Step 4. Isolate glycophorin from the reaction mixture and analyze the protein by digestion and analysis of the peptides.

5.4 Hydropathy plots for prediction of membrane protein topography

---

The number of membrane proteins that have been sequenced is growing and analysis of the collected data has shown that the transmembrane domain of membrane proteins generally consists of 20–25 amino acid residues with a high level of hydrophobicity in an α-helix. With the development of molecular biology techniques it is frequently possible to determine the cDNA sequence and hence the amino acid sequence of proteins. By analysis of the amino acid hydrophobicity and comparing this with known membrane proteins it is possible to predict the presence and number of transmembrane domains. Such analysis is termed a hydropathy (or hydrophobicity) profile (Box 5.16).

It must be emphasized that hydrophobic analysis is a predictive technique and should be followed up by experimental appraches such as those described above. However, this does provide a very useful first approach to determining the topography of membrane proteins. Hydropathy plots for the reaction centre of *Rhodobacter sphaeroides* were available before the structure was elucidated. However, these correctly predicted the number and position of the transmembrane regions. This gives us confidence in the predictive method.

Box 5.16

Aim: Determination of hydropathy plots for proteins of known sequence.

Hydrophobicity scales   **Hydropathy plots**

	A	B
Phe	2.8	3.7
Met	1.9	4.4
Ile	4.5	3.1
Leu	3.8	2.8
Val	4.2	2.5
Cys	2.5	2.0
Trp	−0.9	1.9
Ala	1.8	1.6
Thr	−0.7	1.2
Gly	−0.4	1.0
Ser	−0.8	0.6
Pro	−1.6	−0.2
Tyr	−1.3	−0.7
His	−3.2	−3.0
Glu	−3.5	−4.1
Asn	−3.5	−4.8
Glu	−3.5	−8.2
Cys	−3.5	−8.8
Asp	−3.5	−9.2
Arg	−4.5	−12.3

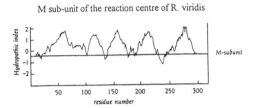

M sub-unit of the reaction centre of R. viridis

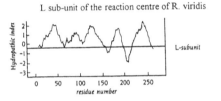

L sub-unit of the reaction centre of R. viridis

EGF receptor

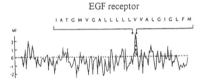

A = Kyle and Dolittle
B = Engelman *et al.*

To prepare a hydropathy plot for a protein the hydrophobic index for each amino acid is calculated. This is the mean hydrophobicity of a window (e.g. 19 amino acids) around the residue. The hydrophobic index is plotted against the position of each amino acid residue to produce a graph describing the hydrophobicity along the length of the polypeptide chain. By comparison with the hydropathy scales of proteins with known transmembrane domains it is possible to predict from this data whether the protein under investigation possesses transmembrane domains. (Adapted from [14].)

5.5 Structural analysis of membrane proteins

In order to understand completely the topography of membrane proteins at a molecular level it is necessary to determine the three-dimensional structure at atomic resolution. This has been achieved by

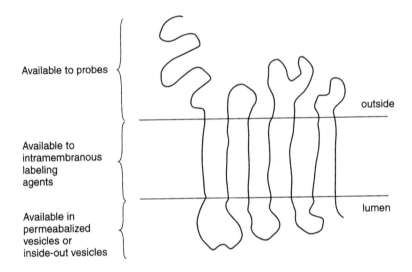

Figure 5.6. Generalized model for probing the topography of membrane proteins.

crystallization and X-ray diffraction of many soluble proteins. However, membrane proteins are difficult to purify and crystallize and this has slowed progress. The first protein for which useful stuctural data were obtained was bacteriorhodopsin. This protein occurs in ordered sheets or two-dimensional crystals in the membranes of *Halobacterium halobium*. High resolution three-dimensional structures for the protein were obtained by analysis of electron micrographs of these membranes [15]. Recently the three-dimensional structure of a second integral membrane protein, the plant light-harvesting complex, was solved by electron microscopy [16]. The first protein membrane crystallized was the photosynthetic reaction center of *R.viridis* [9]. This was achieved by solubilizing the proteins with detergents such as octylglucoside and inclusion of small amphipathic molecules in the crystallization medium. The structures obtained using both X-ray crystallography and electron microscopy for the photosynthetic reaction centers and bacteriorhodopsin are consistent with the general rule that transmembrane domains consist of α-helices of 22–25 amino acid residues. However, the recent crystal structure of bacterial porin has demonstrated the existence of transmembrane β-barrels as a second structural motif [17]. With the development of the methodology for crystallization of membrane proteins and for preparation of two-dimensional crystals of membrane proteins it may be expected that structures of more membrane proteins will emerge in the next few years.

References

1. **Bordier, C.J.** (1981) *J. Biol. Chem.* **256:** 1604–1607.
2. **Cross, G.A.M.** (1990) *Ann Rev. Cell Biol.* **6:** 1–39.
3. **Low, M.G.** (1987) *Biochem. J.* **244:** 1–13.
4. **Marsterson, W.J. and Magee, A.J.** (1991) in *Protein Targetting: A Practical Approach* (A.J. Magee and T. Wileman, eds), p. 223–257. IRL Press at Oxford University Press, Oxford.
5. **Lisani, M.P. and Rodriguez-Boulan E.** (1990) *Trends Biochem. Sci.* **15:** 113–118.
6. **Magee, A.I.** (1990) *J. Cell Sci.* **97:** 581–584.
7. **Chow, M., Der, C.I. and Buss, J.E.** (1992) *Curr. Opin. Cell Biol.* **4:** 629–636.
8. **Casey, P.J.** (1994) *Curr. Opin. Cell Biol.* **6:** 219–229.
9. **Deisenhofer, J., Epp, O., Miki, K., Huber, R. and Michel, H.** (1985) *Nature* **316:** 618–624.
10. **Muller, W.A.** (1994) in *Biomembrane Protocols II* (J.M. Graham and J.A. Higgins, eds), p. 19–42. Humana Press, Totowa, NJ.
11. **Wilkinson, J., Higgins, J.A., Groot, P.H.E., Gherardi, E. and Bowyer D.** (1993) *J. Lipid Res.* **38:** 122–136.
12. **Kahane, I. and Gitler, C.** (1987) *Science* **201:** 351–352.
13. **Ovchinnikov, Y.A.** (1987) *TIBS* **12:** 434–438.
14. **Branden, C. and Tooze, J.** (1991) *Introduction to Protein Structure.* Garland Publishing, New York.
15. **Henderson, R., Baldwin, J.M., Ceska, T.A., Zemlin, F., Beckmann, E. and Downing, K.H.** (1990) *J. Mol. Biol.* **213:** 899–929.
16. **Kuhlbrant, W., Wang, D.N. and Fujioshi, Y.** (1994) *Nature* **367:** 614–621.
17. **Kreusch, A., Neubuser, A., Schiltz, E., Weckesser, J. and Schultz, G.E.** (1994) *Protein Sci.* **3:** 58–63.

6 Investigation of membrane lipids

The major lipid components of biological membranes consist of phospholipids, and, in eukarytoes, a variable amount of a sterol (see Chapters 1 and 2). The lipids are arranged in the form of a bilayer, although there is evidence that transient nonbilayer hexagonal arrangements may occur. The phospholipid bilayer is asymmetric. However, unlike proteins, membrane lipids do not have an absolute asymmetry; rather the two halves of the bilayer contain the same phospholipids, but differ in percentage composition.

6.1 Determination of the organization of membrane phospholipids

Phospholipids are amphipathic molecues, which aggregate when they are dispersed in water. The most thermodynamically stable arrangement is adopted. Essentially, the hydrophobic parts (acyl chains) of the lipid molecules associate to exclude water, while the hydrophilic parts orient towards the aqueous medium. Phospholipids adopt several arrangements, depending largely on the shape of the molecules. Thus, cylindrical molecules (e.g. phosphatidylcholine) form bilayers, cone-shaped molecules (e.g. phosphatidylethanolamine) form hexagonal HII arrangements, and inverted cones (e.g. lysophospholipids) form micelles (*Figure 6.1*). These structures can be distinguished using physical techniques, X-ray diffraction, freeze-fracture electron microscopy, or ^{31}P NMR [1, 2]. The same techniques have been used to study the organization of lipids in biological membranes. Although these contain a more complex mixture of phospholipids, the evidence suggests that the bilayer is the predominant form. However, there is evidence, from NMR studies, that the hexagonal HII form can also occur.

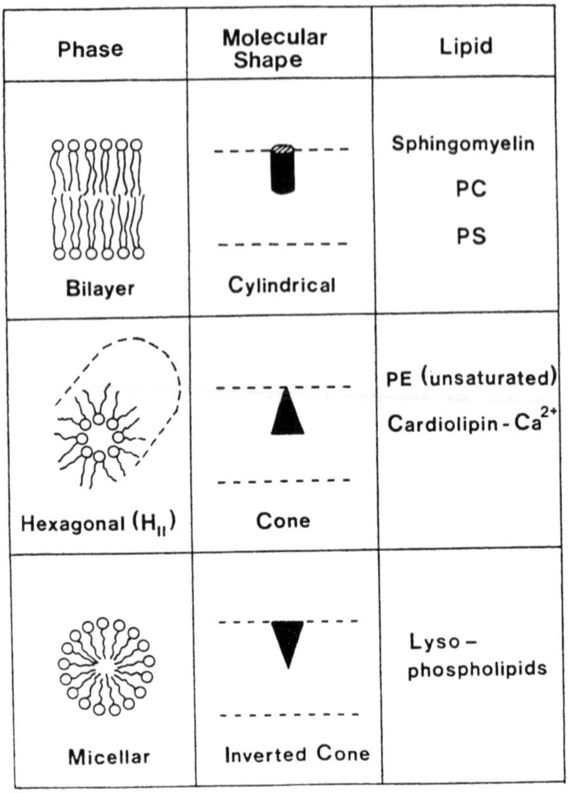

Phase	Molecular Shape	Lipid
Bilayer	Cylindrical	Sphingomyelin PC PS
Hexagonal (H$_{II}$)	Cone	PE (unsaturated) Cardiolipin - Ca^{2+}
Micellar	Inverted Cone	Lyso – phospholipids

Figure 6.1. Structures adopted by phospholipids in aqueous dispersion. PC, phosphatidylcholine; PS, phosphatidylserine; PE, phosphatidylethanolamine.

6.2 Model membranes

The physical properties and functions of the different lipid species of membranes have been studied in model membranes, which are simpler than biological membranes [2]. Reconstituted membranes, containing a single purified protein in a defined lipid bilayer, have also been used to investigate membrane function.

6.2.1 Multilamellar vesicles (MLVs)

These are formed by drying a solution of phospholipid in a film on the inner surface of a suitable glass vessel, followed by rehydration of the film in a buffer using mechanical methods such as vigorous mixing on a vortex mixer. This method produces structures made up of a series of concentric phospholipid bilayers. MLVs have been used in investigations of the physical characteristics of membrane lipids. However, because only a small fraction of the lipid is in the outer layer, MLVs have limited value in other studies of membrane function.

6.2.2 Small unilamellar vesicles (SUVs)

These are made by the passage of MLVs through a French pressure cell or by sonication of MLVs. They consist of small vesicles with diameters of 25–40 nm. Because of the small size, the bilayer of SUVs has a high degree of curvature.

6.2.3 Large unilamellar vesicles (LUVs)

These consist of single bilayer vesicles of 50–500 nm diameter. They are prepared in several ways:

- The lipids are dissolved in detergent. The detergent used should have a high critical micellar concentration so that it can subsequently be removed by dialysis or gel filtration (e.g. octylglucoside or cholate). As the detergent concentration decreases, the lipids form LUVs.
- The lipid is dissolved in an organic solvent such as ether or ethanol. This is slowly injected into an aqueous buffer. As the solvent is diluted, LUVs form.
- MLVs are extruded under pressure through polycarbonate filters of defined pore size [3]. This technique can be used to prepare LUVs of uniform diameter regulated by the pore size of the filters. This is the method of choice if the experiment in hand precludes the use of solvents or detergents.

6.2.4 Black lipid membranes

These are used most frequently by electrophysiologists to study current flow across membranes. They are prepared by painting a film of phospholipid in solution in a hydrocarbon across a small aperture (about 2 nm in diameter) in a sheet of material, which separates two compartments filled with buffers. Electrodes are placed in the compartments to measure the electrical properties of the bilayer. These membranes can include a protein and have been used to investigate ion channels

6.3 Reconstituted membranes

A variety of membrane proteins have been reconstituted into membrane vesicles [4] (see Box 6.1).

Box 6.1

Aim: To reconstitute a membrane protein (Ca^{2+}/Mg^{2+}-ATPase) into lipid vesicles.

Protocol

Step 1. Purify the ATPase from sarcoplasmic reticulum [6] and solubilize it in buffered cholate solution.

Step 2. Dissolve phosphatidylethanolamine and phosphatidylcholine (4:1, mol/mol) in chloroform/methanol, 2:1, v/v. Dry the lipid in a thin film in a glass tube. Disperse the lipid in buffered cholate by sonication.

Step 3. Mix lipid and ATPase solutions (see General applicability below) and remove the detergent by passing the mixture through a Sephadex G-50 column or by dialysis against an appropriate buffer.

General applicability: Depending on the ratio of protein to lipid used, these procedures yield a heterogeneous mixture of vesicles, some of which will contain several protein molecules with different orientations. For many studies asymmetric vesicles are required. To achieve this, low protein to lipid ratios (as low as 1 : 5000, w/w) are used. This produces vesicles containing one or no protein molecules. Those containing the protein in the desired orientation can then be separated by affinity chromatography.

6.4 Determination of the transverse distribution of membrane phospholipids

The basic principles involved in probing the transverse distribution of membrane lipids are simple. Lipids that are available for modification by probes are in the outer leaflet of the membrane bilayer, and those only available in permeabilized vesicles are in the inner leaflet. The experimental design and controls are similar to those for determining the orientation of membrane proteins (see Chapter 5).

- The membrane vesicles investigated must be closed, of the same orientation, and derived from a single type of organelle.
- The probe used should react with the membrane lipids in a detectable way and not cause rearrangement of the membrane bilayer.

- Control experiments should demonstrate that the vesicles do not leak during probing experiments and that the probe does not penetrate the vesicles.

Among the probes used to determine the transverse distribution of membrane phospholipids are:

- chemical labeling reagents;
- phospholipases;
- phospholipid exchange proteins;
- determination of the surface exposure of phosphatidylserine using an assay based on prothrombinase.

The first two methods are invasive, as they modify the phospholipid (markedly in the case of the enzymic methods). The second two methods are less invasive and, in theory, should not disturb the membrane organization.

6.4.1 Chemical probes

The chemical probes available for reaction with membrane phospholipids are largely restricted to those that react with amino groups in phosphatidylethanolamine and phosphatidylserine. Two examples are shown in Boxes 6.2 and 6.3. The first uses trinitrobenzene sulfonate (TNBS) which reacts with NH_2 groups (*Figure 6.2*) to form stable products which are detected by their yellow color; the second is based on the use of two radiolabeled probes, one of which penetrates the membrane, while the second remains outside.

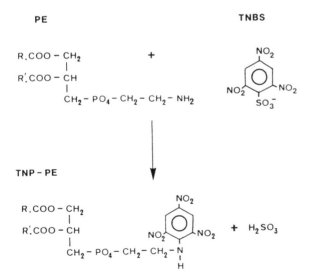

Figure 6.2. Reaction of aminophospholipids with TNBS.

These are detected by autoradiography, image analysis, phosphor-imaging or by counting the lipid bands in a scintillation counter.

6.4.2 Phospholipases

A variety of phospholipases of different specificity are available from commercial sources. These include phospholipase C, phospholipase D, phospholipase A1 and A2, sphingomyelinase C, and the phosphatidylinositol-specific phospholipase C (*Figure 6.3*). The details for the use of each enzyme probe are generally provided by the supplier. If not, preliminary experiments should be carried out using phospholipid dispersions to determine the amount of enzyme needed and the incubation conditions.

Box 6.2

Aim: To determine the proportion of membrane aminophospholipids in the outer leaflet of microsomal vesicles [6].

Protocol

Step 1. Prepare endoplasmic reticulum vesicles (microsomes) as in Chapter 3 in an appropriate buffer (this should not contain amino groups).

Step 2. Incubate the microsomes (about 5 mg protein ml^{-1}) with TNBS (3 mM).

Step 3. Stop the reaction with 10% trichloroacetic acid, pellet the denatured microsomes by centrifugation and wash the pellet in distilled water or buffer.

Step 4. Extract the lipids from the pellet and separate the phospholipids by thin layer chromatography (TLC) (see Chapter 4).

Step 5. Outline the yellow TNP-phospholipid bands (which have different Rfs from the unreacted phospholipids) lightly in pencil, before staining all of the phospholipid bands with iodine vapor or spraying the plate with 0.25% ninhydrin, followed by warming. The aminophospholipids stain purple.

Step 6. Scrape the TNP-phosphatidylethanolamine, TNP-phosphatidylserine and the unreacted aminophospholipids and determine the phospholipid phosphorus.

Interpretation: The percentages of recovered phosphatidylethanolamine and phosphatidylserine in the TNB-phospholipid bands are equivalent to those in the outer leaflet of the bilayer.

Controls: Microsomes should be incubated without TNBS to check recovery of the aminophospholipids. The integrity of the vesicles should be checked (see Chapter 5). The experiments should be repeated on open vesicles, for example using sodium carbonate [7] to react with TNBS.

General applicability: This method can be applied to all membrane vesicles and to isolated cells provided suitable controls are performed (see Chapter 5).

Box 6.3

Aim: To determine the proportion of aminophospholipids in the outer leaflet of erythrocyte membranes.

Principle: Ethylacetimidate (EA) reacts with free amino groups and does not penetrate membranes. Isoethionylacetimidate (IEA) also reacts with free amino groups but does penetrate membranes.

Protocol

Step 1. Incubate washed erythrocytes with ^{14}C-EA or ^{14}C-IEA.

Step 2. Pellet the cells by centrifugation and wash.

Step 3. Extract the lipids and separate the phospholipids on by TLC (see Chapter 4).

Step 4. Scrape the lipid bands, add scintillation fluid and count in a beta counter. Or detect and count the radiolabeled bands using an image analyzer.

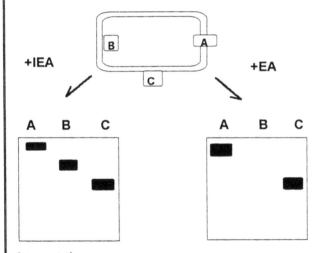

Interpretation
Phospholipid A is in both leaflets of the bilayer. The percentage in each leaflet can be calculated from the distribution of radioactivity with the two labels.
Phospholipid B is in only the inner leaflet.
Phospholipid C is in only the outer leaflet; i.e. the d.p.m. in the band are the same with both radiolabels.

Experiments in which phospholipases are used as probes involve the following steps:

- incubation of the membrane vesicles with the enzyme;
- extraction of the lipids and separation and determination of the phospholipids by TLC (see Chapter 4);
- calculation of the amount of phospholipid hydrolyzed (i.e. lost from the membranes) by enzyme treatment, by comparison with untreated membranes.

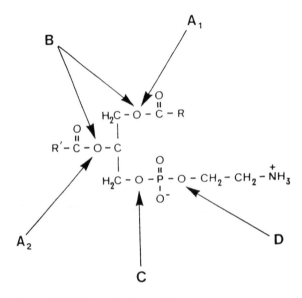

Figure 6.3. Sites of action of phospholipases (see text for further details).

As with all probing experiments, it is necessary to demonstrate that the vesicles remain closed during the probing experiments. It is also necessary to use opened vesicles to demonstrate that lack of hydrolysis is not due to the enzyme's specificity (see Chapter 5 and above).

6.4.3 Phospholipid transfer proteins

Phospholipid transfer proteins (PL-TPs) have been isolated from a number of tissues, including heart, liver and brain [8, 9]. These proteins catalyze a one for one exchange of phospholipid molecules between membranes, lipoproteins or liposomes. Since PL-TPs only exchange the phospholipids of the outer leaflet of membrane bilayers and do not appear to alter the structure of membranes, they provide a unique tool for investigating the transverse distribution of membrane phospholipids (Box 6.4). The PL-TPs fall into three categories depending on their specificity:

- those specific for phosphatidylcholine;
- those with high activity for phosphatidylinositol;
- those with broad specificity for all phospholipid classes.

Preparation of radiolabeled donor membranes. If the membranes under investigation are to be prepared from animal tissue, such as rat liver, the phospholipids are labelled *in vivo* by intraperitoneal or intravenous injection of an appropriate radiolabeled precursor (e.g. ^{3}H-glycerol, ^{3}H-palmitic acid, ^{3}H-oleic acid, ^{3}H-choline or ^{3}H-ethanolamine). If cells in culture or subcellular fractions from these are used the phospholipids are labeled by addition of the radiolabeled precursor to the incubation medium. Preliminary experiments should be carried out to determine the amount of isotope required to obtain sufficient incorporation into the phospholipids of interest. In general, ^{3}H-labelled substrates are considerably less expensive than those labelled with ^{14}C, and are therefore preferable for *in vivo* radiolabeling.

Box 6.4

Aim: To probe the transverse distribution of membrane phospholipids using PL-TP.

Protocol

Step 1. Incubate the radiolabeled membranes (donor membrances) with an excess of liposomes (acceptor membranes) and PL-TP.

Step 2. Separate the membranes and liposomes by centrifugation on a sucrose cushion.

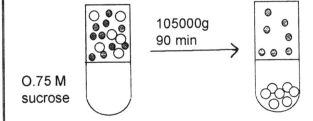

105000g
90 min

0.75 M
sucrose

Step 3. Extract the phospholipids from the pelleted membranes and from the liposome suspension and determine the transfer of radiolabel.

Interpretation: It is assumed that only the phospholipid in the outer leaflet of acceptor and donor species is exchanged. See text for the calculation.

Controls: Preliminary experiments should be carried out without PL-TP to ensure that the the donor and acceptor vesicles are complete.

Preparation of the acceptor membranes. A variety of accceptor membranes can be used provided that these can be separated from the donor membranes. Liposomes are generally lighter than biological membranes and have been used frequently. However, some investigators have also used mitochondria or plasma lipoproteins. To prepare liposomes containing phosphatidylcholine the following steps should be followed:

- Mix phosphatidylcholine and phosphatidic acid (98:1, w/w) in chloroform/methanol, 2:1, v/v. Dry the lipid in a thin film on the sides of a glass tube.
- Add buffer and sonicate to disperse lipids.
- Centrifuge the mixture in an ultracentrifuge at approx. 100 000g for 90 min to pellet particulate lipid. The supernatant contains the liposomes.

Calculation. It is assumed that phospholipids are labeled uniformly in the acceptor molecules and that two-thirds of the liposome phosphatidylcholine is in the outer leaflet of the liposome bilayer. On this basis:

$$\text{d.p.m. transferred} = (\text{PC-lipo/PC-lipo} + \text{PC-donor}) \times (\text{PC-o/PC-donor}) \times \text{total d.p.m.}$$

where: PC-lipo is the exchangeable pool of phosphatidylcholine in the liposomes (this is assumed to be two-thirds of the total in SUV); PC-donor is the total pool of PC in the membranes; PC-o is the pool of PC in the outer leaflet; d.p.m. transferred is the proportion of the d.p.m. in the acceptor at the end of the incubation; and total d.p.m. is the initial d.p.m. in the donor.

6.4.4 Assay based on prothrombinase

The coagulation cascade is a complex series of reactions involving a number of factors. *In vivo* transfer of phosphatidylserine to the outer leaflet of platelets acts as a catalyst for thrombin activation. Advantage has been taken of this to design a method for assay of the amount of phosphatidylserine exposed at the outer surface of membrane vesicles [10]. The incubation is set up so that all reagents are present in excess. The amount of thrombin activity is then equivalent to the amount of phosphatidylserine exposed at the outer leaflet of the membrane vesicle (Box 6.5).

Box 6.5

Aim: To determine the amount of phosphatidylserine exposed at the outer leaflet of membrane vesicles.

Principle: Phosphatidylserine in a membrane acts as a catalyst for the coagulation cascade. Factor Xa + Factor Va bind to the phosphatidylserine-containing membranes. This activates the complex which catalyzes the conversion of prothrombin to thrombin. Thrombin is then assayed colorimetrically. Provided all reactants are present in excess the amount of thrombin is equivalent to the amount of phosphatidylserine exposed at the membrane surface. The amount of phosphatidylserine in the membranes is determined by lipid analysis and from the above assay the proportion exposed at the outer surface can be determined.

6.5 Determination of the transverse distribution of membrane cholesterol

The amount of cholesterol in the outer leaflet of membrane vesicles or cell membranes can be determined by its accessibility to cholesterol oxidase. Cholesterol is converted to cholestenone, which is determined spectrophotometrically at 235 nm [11]. Such studies have shown that more than 90% of the cholesterol is available for oxidation under conditions in which the cells or membrane vesicles remain sealed. Apparently cholesterol rapidly moves from one leaflet of the membrane bilayer to the other, with a half-life of less than 1 min [12].

6.6 Studies of lipid transport

With few exceptions, membrane phospholipids are synthesized in the endoplasmic reticulum by enzymes located at the cytosolic side of this membrane [13, 14]. Intracellular transport of phospholipids is extremely complex (see [14]). Lipids move across bilayers and between membranes in a forward (i.e. from the endoplasmic reticulum to the plasma membrane) and backward direction. Lipids are also rearranged in relation to function. Transport of lipids includes:

- Transbilayer movement of newly synthesized lipids from the cytosolic leaflet to the lumenal leaflet of the endoplasmic reticulum.
- Movement of lipid between membranes. The mechanisms implicated in this process include: (i) transfer as a component of membrane vesicles; (ii) molecular transfer by PL-TP; and (iii) transfer by membrane fusion.
- ATP dependent protein-transporter catalyzed transbilayer movement of aminophospholipids in the plasma membrane.

6.6.1 Investigation of the transmembrane movement of phospholipids

The rate of transmembrane transfer of phospholipids can be determined by labeling lipid in one leaflet of the bilayer and following its transfer to or from the other leaflet by taking advantage of the methods described above for determining the distribution of phospholipids. For such studies phospholipids have been prepared incorporating radiolabels, tagged with spin-labeled molecules (*Figure 6.4*) and tagged with fluorescent molecules (*Figure 6.5*) .

Investigation of the transbilayer movement of phospholipids using radiolabels. This includes the following steps:

- Incorporation of radiolabeled precursor into the phospholipid in one leaflet. This can be achieved in the endoplasmic reticulum by incubating the isolated vesicles with an appropriate radiolabeled substrate (e.g. CDP-^{3}H-choline, CDP-^{3}H-ethanolamine, ^{3}H-acyl-CoA). As the enzymes involved in synthesis of phospholipids are on the cytosolic side of the bilayer, initial incorporation is at this site. To label the plasma membrane, cells have been incubated with liposomes containing the radiolabeled lipid.
- Determination of the amount of labeled phospholipid available to a probe (covalent label, phospholipases or PL-TP) which cannot penetrate the vesicle or cell membrane. This is carried out at sequential times, and the movement of phospholipid to or from the accessible pool determined. This is related to the rate of transbilayer movement.
- Carry out appropriate controls as described above (Section 6.6).

The prothrombinase assay for determination of the rate of transmembrane movement of phosphatidylserine. As indicated above, phosphatidylserine exposed at the surface of cells acts as a catalyst for activation of prothrombinase. Thus, the rate of transfer of phosphatidylserine to the outer leaflet of isolated membrane vesicles,

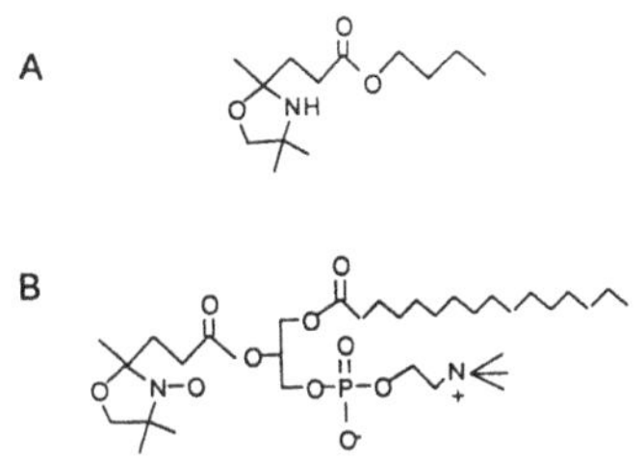

A

B

Figure 6.4. Structure of spin-labeled lipids. (a) Spin-labeled fatty acid with the spin-label attached to the COOH; (b) spin-labeled PC labeled in the polar head group.

or cells in suspension or culture, can be determined by measuring the amount of this lipid exposed at the surface of the platelets at timed intervals.

The use of spin-labeled molecules to determine the rate of transmembrane movement of phospholipids. Pioneer studies by Kornberg and McConnell [15] used spin-labeled phospholipids to determine the rate of transmembrane movement of phospholipids in model membranes. Phosphatidylcholine with a nitroxide group

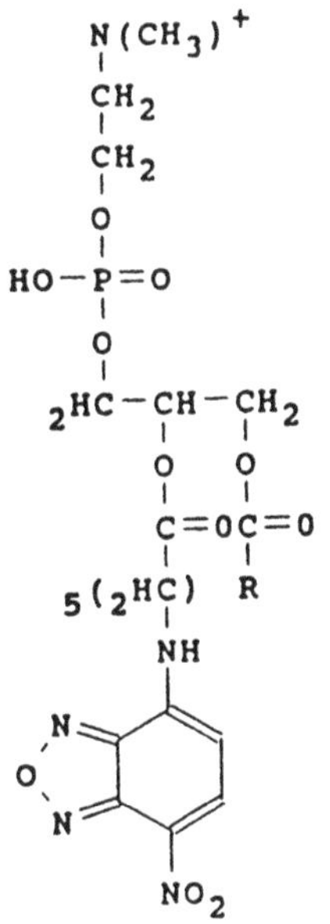

$N(CH_3)^+$
|
CH_2
|
CH_2
|
O
|
$HO-P=O$
|
O
|
$_2HC-CH-CH_2$
| |
O O
| |
$C=OC=O$
|
$5(_2HC)$ R
|
NH

NO_2

Figure 6.5. Structure of NBD-labeled PC.

Box 6.6

Use of spin-labeled phosphatidylcholine to determine the rate of transmembrane movement of phospholipids in model membranes.

Protocol: Prepare SUV containing spin-labeled phospholipid mixed with unlabeled phospholipid.

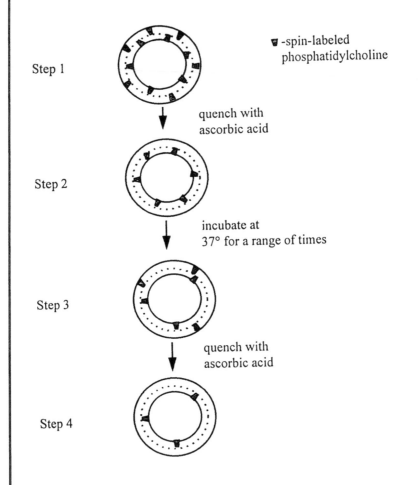

Step 1

▼ -spin-labeled phosphatidylcholine

quench with ascorbic acid

Step 2

incubate at 37° for a range of times

Step 3

quench with ascorbic acid

Step 4

Measurement of the concentration of the spin-labeled phospholipid at Step 2 and Step 4 by esr allows determination of the rate of transmembrane movement.

incorporated into the polar head group was used in these studies. A small amount of this phospholipid was mixed with egg lecithin and was dispersed in liposomes. Ascorbic acid, a nonpermeant probe, was used to abolish the paramagnetism of the spin-labeled molecules exposed at the outer surface of the vesicles. By subsequent exposure to ascorbate it was possible to determine the rate of transfer of the spin-labeled phospholipid from the inner leaflet to the outer leaflet of the vesicle bilayer by measuring changes in its electron spin resonance spectrum (e.s.r.) (Box 6.6). A half-life for transfer of 6.5 h at 30°C was found. Similar rates of transfer in model membranes and biological membranes have been obtained by other laboratories using different approaches. Therefore the nitroxide probe does not modify the rate of phospholipid transfer.

More recent studies have shown that specific membrane proteins act as carriers for ATP-dependent transmembrane movement of aminophospholipids. However, the nitroxide in the polar head group interferes with the carrier protein and in order to investigate carrier-mediated phospholipid transfer it is necessary to incorporate the spin-label elsewhere in the molecule, usually in the acyl group (*Figure 6.4*) [15]. As this spin-label is not available to ascorbate in the incubation medium, the amount of label in the outer leaflet is determined by back-exchange on to albumin (Box 6.7). In contrast to experiments using the prothrombinase assay, analogs of all major membrane phospholipids can be synthesized [16].

Box 6.7

Aim: Determination of the rate of transbilayer movement of spin-labeled phospholipids in erythrocyte membranes.

Protocol: To determine transfer from the outer leaflet to the inner leaflet.

Step 1. Incubate erythrocyte suspension with a suspension of the nitroxide-labeled phospholipid dispersed in buffer.

Step 2. Take aliquots of the suspension immediately and at timed intervals and transfer to solutions of bovine serum albumin on ice. After 1 min pellet the cells by centrifugation.

Step 3. Determine the amount of spin-label in the bovine serum albumin-containing supernatant from its e.s.r. spectrum [16].

Interpretation: The amount of spin-label recovered in the supernatants represents the amount of labeled phospholipid in the outer leaflet at the time of sampling. From this data the rate of transmembrane movement of phospholipid can be calculated.

General applicability: This method can be used for any cell type. Movement of the label into the cells from the outer leaflet can also be determined using this approach.

6.6.2 *Transport of phospholipids and cholesterol between intracellular membranes*

Transfer of phospholipids (or cholesterol) can be followed using labeling methods described above combined with cell fractionation or microscopy in the case of fluorescently labeled lipids.

Transfer of newly synthesized phospholipids from the endoplasmic reticulum to other membranes. This process may be investigated by a combination of pulse-radiolabeling and cell fractionation.

- Incubate cells in suspension with a radiolabeled precursor or inject the precursor intravenously to label whole tissue, such as liver.
- Homogenize the cells or tissues and prepare subcellular fractions (see Chapter 3) at a range of times after incorporation of the radiolabel.
- Extract the phospholipids, separate them by TLC and analyze the d.p.m. incorporated and the specific activities of the phospholipid classes.
- From this data the rate of transport of lipids can be calculated. A similar approach can be used to follow cholesterol transport.

Investigation of the intracellular transfer of phospholipids using fluorescent analogs. A novel method for investigation of phospholipid transport was developed by Pagano and co-workers [17], who synthesized a range of phospholipid analogs with fluorescent tags, the most common being nitrobenzoxadiazole (NBD) (*Figure 6.5*) [18]. NBD-labeled lipids are incorporated into liposomes and spontaneously transfer into the plasma membrane of cells. The excess liposomes are removed by washing. This results in pulse-labeling the cells with the fluorescent marker, the intracellular movement of which can be followed by fluorescence microscopy. This method can also be used to follow the internalization of the spin-label by determining how rapidly the labeled lipid moves from a pool, which can be removed by incubation with liposomes. This method has also been used to determine the characteristics of the aminophospholipid transporters.

6.7 Genetic approach to investigate membrane phospholipids

Genetic studies of mutants, particularly in yeasts, have provided a complementary approach to biochemical studies in elucidating factors involved in regulation of protein sorting (see Chapter 7). Only recently have such approaches been successful in studies of the regulation of the intracellular transit of phospholipids. However, this is a promising approach which should provide new information in the future [19].

References

1. **Cullis, P.R. and De Kruijff, B.** (1979) *Biochim. Biophys. Act* **559:** 399–420.
2. **Cullis, P.R. and Hope, M.J.** (1991) in *New Comprehensive Biochemistry, Vol. 20: Biochemistry of Lipids, Lipoproteins and Membranes* (D.E. Vance and J.E. Vance, eds), pp. 1–40. Elsevier Science, Amsterdam.
3. **Hope, M.J., Bally, M.B., Mayer, L.D., Janoff, A.S. and Cullis, P.R.** (1986) *Chem. Phys. Lipids* **40:** 89–108.
4. **Maden, T.D.** (1986) *Chem. Phys. Lipids* **40:** 207–236.
5. **East, J.M.** (1994) in *Methods in Molecular Biology, Vol. 27: Biomembrane Protocols II. Architecture and Function* (J. Graham and J. Higgins, eds), 87–95. Humana Press, Totowa, NJ.
6. **Higgins, J.A. and Piggott, C.A.** (1992) *Biochim. Biophys. Acta* **693:** 151–158.
7. **Fujiki, Y., Hubbard, A.L., Fowler, S. and Lazarow, P.B.** (1982) *J. Cell. Biol.* **93:** 97–102.
8. **Wirtz, K.W.A.** (1991) *Annu. Rev. Biochem.* **60:** 73–99.
9. **Ossendorf, B.C., Snoek, G.T. and Wirtz, K.W.A.** (1994) *Current Topics Membranes* **40:** 217–259.
10. **Comfurius, P., Bevers, P. and Zwaal, R.F.A.** (1994) in *Methods in Molecular Biology, Vol. 27: Biomembrane Protocols II. Architecture and Function* (J. Graham and J. Higgins, eds), pp.131–142. Humana Press, Totowa, NJ.

11. **Lange, Y., Dolde, J. and Steck, T. L.** (1991) *J. Biol. Chem.* **256:** 5321–5323.
12. **Lange, Y. and Ramos, B.V.** (1983) *J. Biol. Chem.* **258:** 15130–15134.
13. **Bishop, W.R. and Bell, R.M.** (1988) *Annu. Rev. Cell Biol.* **4:** 575–610.
14. **Moreau, P. and Cassagne, C.** (1994) *Biochim. Biophys. Acta* **1197:** 257–290.
15. **Kornberg, R.D. and McConnell, H.M.** (1971) *Biochemistry* **10:** 1111–1120.
16. **Fellman, P., Zachiwski, A. and Devaux, P.** (1994) in *Methods in Molecular Biology, Vol. 27: Biomembrane Protocols II. Architecture and Function* (J. Graham and J. Higgins, eds), pp. 161–176. Humana Press, Totowa, NJ.
17. **Pagano, R.E. and Sleight, R.G.** (1985) *Science* **229:** 1051–1057.
18. **Sleight, R.** (1994) in *Methods in Molecular Biology, Vol. 27: Biomembrane Protocols II. Architecture and Function* (J. Graham and J. Higgins, eds), pp. 143–153. Humana Press, Totowa, NJ.
19. **Alb, J.G., Kearns, M.A. and Bankaitas, V.A.** (1996) *Curr. Opin. Cell Biol.* **8:** 534–541.

7 Protein targetting and membrane biogenesis

Eukaryotic cells contain a large number of membrane compartments many of which have distinct domains (see Chapter 2). Each of these has characteristic protein components. All protein synthesis is initiated by cytosolic ribosomes. Therefore cells have a major problem in sorting and targetting of newly synthesized proteins, so that they are delivered to the correct destination (*Figure 7.1*). In addition, cells are not static structures. There is a continuous movement of membrane from the ER to the plasma membrane, via the Golgi during secretion, and movement of plasma membrane vesicles into the cell in endocytosis. Despite this interchange, the membranes in each compartment retain their characteristic composition. Therefore mechanisms must exist for protein retention as well as delivery to the correct location. There are numerous studies of the molecular details and mechanisms involved in protein targetting in prokaryotes and eukaryotes. In general the techniques used are those of molecular biology. In this chapter the ways in which protein targetting can be investigated will be outlined, with selected examples to illustrate the range of experimental approaches which can be used. Such approaches are transferrable and can be used to investigate a variety of membranes.

7.1 Genetic approaches to studies of protein targetting

Classical genetic methods have been used to investigate protein transport in eukaryote cells. Using this approach, mutant cells deficient in secretion or protein targetting are isolated and the defective genes identified. A collection of genes termed *sec*, which function at different stages of the secretory pathway, has been identified in the budding yeast *Saccharomyces cerevisae* [1] and also in the fission yeast *Saccharomyces pombe* [2]. Yeasts have also been

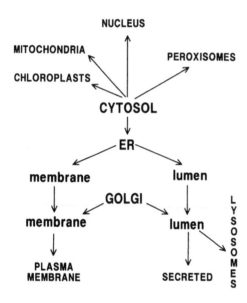

Figure 7.1. Destination of newly synthesized proteins.

used to investigate membrane protein targetting [3–5]. In many cases, identification of the gene products in the deficient mutants has shown that these are analogous to proteins identified using biochemical methods in other eukaryotic cells, and it is clear that the general characteristics of membrane traffic are conserved between yeasts and higher organisms. The genetic approach is a very powerful one. However, it does require the ability to prepare mutants and to identify their phenotypes, and has largely been restricted to organisms in which the genes are easily manipulated. A detailed account of the specialized methods used is outside the scope of this book; however, for such methods the reader is referred to protocols published elsewhere [6,7].

7.2 Biochemical approaches to studies of protein targetting

Protein molecules carry the information which directs them to their final environment. This 'address tag' can reside in a specific amino acid sequence or in some part of the structure of the folded protein. The aim of the biochemical studies is to identify the nature of these signals. Despite the complexity of the pathways involved, the techniques used in such studies are relatively few. These include:

- **Cell-free systems** in which mRNA for the protein under investigation is translated in a cell-free system which contains all of the necessary co-factors prepared from reticulocytes or wheat-germ lysates. In most cases the proteins are labeled by including radiolabeled amino acids in the medium. Putative factors or membranes are added to the system to determine whether they are involved in targetting of the newly synthesized protein.
- **Use of cDNA constructs** prepared by combining the nucleotide sequence for a known protein with putative targetting sequences. The mRNAs are translated in cell-free systems or cDNA incorporated into plasmid vectors and transfected into cells in culture. The protein is detected in the cells using specific antibodies linked to fluorescent markers for light microscopy, or electron-dense markers (e.g. horseradish peroxidase or gold particles) for electron microscopy. Alternatively, the newly synthesized protein is detected by immunoblotting (see Chapter 4).
- **Site-directed mutagenesis** of specific amino acids in putative signal sequences. The specificity of these can be determined by using the modified cDNA in transfection experiments as in the point above.
- **Data base searches** for common sequences or structural motifs in proteins of known location.
- **Reconstituted membranes** in which purified proteins are incorporated into liposomes to investigate the characterisitics of the targetting and translocation systems (see Chapter 6).

7.2.1 Steps in protein targetting

Despite the great complexity in membrane targets for proteins there are a number of steps which are generally common to the overall process:

- Protein synthesis is initiated in the cytosol and is targetted post-translationally to organelles other than the ER or co-translationally to the ER and beyond (*Figure 7.1*).
- Molecular chaperones may stabilize the newly synthesized proteins in a loosely folded state (or arrest translation in the case of proteins targetted co-translationallly to the ER).
- The proteins contact a receptor in the membrane indirectly or directly.
- Insertion of the proteins into the membrane and translocation involve a complex of proteins forming a translocation pore, through which the unfolded protein moves.
- Unidirectional transport of the protein across the membrane may involve molecular chaperones on the inside of the membrane.

- The translocated protein folds with the assistance of chaperones.

A large number of chaperone proteins have now been identified [8–12]. Some were identified using genetic techniques others using biochemical methods. The nomenclature of the chaperones is complicated by the fact that they were discovered by different investigators using different experimental approaches; for example, the heat shock proteins and the glucose regulatory proteins have proved to be the same and these are analogous to a number of mutant proteins (*Table 7.1*).

Table 7.1. The main classes of molecular chaperones involved in protein targetting together with some examples[a]

Class	Location	Names	Main function
hsp-70	*E. coli*	DnaK	Protein stabilization, oligomerization
	Eukaryotic cytosol	Hsc-73, hsp-70	Protein translocation into organelles
	Mitochondria	SSC1	Protein translocation
	Endoplasmic reticulum lumen	BiP, Grp-78	Protein translocation and assembly
hsp-60	Prokaryotic cytosol	GroEL	Protein folding, phage assembly
	Mitochondria matrix	hsp-60	Protein translocation and targetting
	Chloroplasts	Cpn-60	Protein translocation and assembly
SRP	Eukaryotic cytoplasm	Signal recognition particle (SRP)	Targets proteins to endoplasmic reticulum, arrests translation
	E. coli	Sec B, Sec A, Sec Y	Arrests translation, targets proteins to membranes

[a] This is not an exhaustive list.

7.2.2 Nuclear targetting

As described in Chapter 2, the nucleus is surrounded by a double membrane which is penetrated by nuclear pores. These are filled with a protein pore complex [13]. The many proteins which move into or out of the nucleus must cross this pore complex, which allows slow diffusion of proteins of molecular weight above 40 000 kDa and completely blocks those above 60 000 kDa. For entry into the nucleus, proteins must carry a nuclear localization sequence (NLS). Evidence for the existence of a NLS in nucleoplasmin (MW 160 000 kDa, a pentameric protein) includes the following observations:

- Radiolabeled nucleoplasmin (prepared from cells incubated with radiolabeled amino acids) was injected into the cytoplasm of frog oocytes. The radiolabeled protein detected by autoradiography moved into the nucleus. *Therefore nucleoplasmin has a NLS.*

- Removal of the –COOH tail abolished transport into the nucleus. *Therefore the NLS resides in the –COOH tail.*
- The truncated nucleoplasmin injected into the nucleus did not diffuse out. *Therefore the NLS is not necessary for nuclear retention.*
- Gold-labeled nucleoplasmin was seen at the nuclear pore by electron microscopy. *Therefore the nuclear pore is the gate.*
- Construction of cDNA with the –COOH NLS of nucleoplasmin added to pyruvate kinase (a cytosolic enzyme) transfected into oocytes resulted in synthesis of pyruvate kinase which was transported into the nucleus. *Therefore the NLS contains all of the information for nuclear transport.*

A variety of NLSs have been identified (Box 7.1). These can vary in the position of the NLS in the molecule and in its amino acid sequence. The common feature of the NLS is a high concentration of basic amino acids.

Box 7.1

Examples of NLS: sequence and position in the protein

Nucleoplasmin	COOH end	Lys Arg X X X X Lys X X X X (Lys)4
SV 40 T antigen	residues 126–132	(Lys)3 Arg Lys Val
Yeast ribosomal protein	residues 1–21	Pro Arg Lys Arg

Transport of proteins into the nucleus involves two steps: binding to the nuclear pore complex and transport into the nucleoplasm. The latter step requires energy and is dependent on cytosolic factors. See Box 7.2 and Box 7.3 for examples of ways of studying import into the nucleus. In either case the specific requirements for uptake and transport can be investigated by addition of various putative factors or by altering the structure of the cDNA or the protein investigated.

7.2.3 Targetting to peroxisomes

Peroxisomal lumenal enzymes, which include a range of oxidative enzymes (see Chapter 2), are synthesized in the cytosol and targetted to the peroxisome post-translationally. The targetting sequence for peroxisomal enzymes was identified by comparison of the amino acid sequences of the peroxisomal enzyme which had been deposited in data bases. Luciferase, catalase D-amino acid oxidase, acyl CoA

Box 7.2

Aim: To determine factors involved in transport of proteins into nuclei.

Protocol

Step 1. Prepare fluorescently labeled protein with NLS or conjugate NLS to a protein such as albumin.

Step 2. Prepare isolated nuclei.

Step 3. Incubate nuclei and labeled protein with putative factors involved in nuclear transport. Examine nuclei under light microscope. Determine whether the protein binds or is transported into the nucleus.

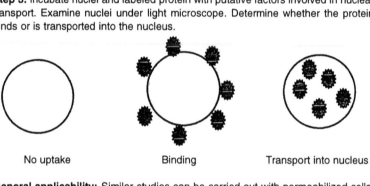

| No uptake | Binding | Transport into nucleus |

General applicability: Similar studies can be carried out with permeabilized cells. A confocal microscope allows clear identification of binding compared with uptake.

Box 7.3

Aim: To determine whether a specific protein is targetted to the nucleus.

Protocol

Step 1. Prepare a plasmid vector in which the cDNA for the protein is inserted.

Step 2. Transfect tissue culture cells with the vector.

Step 3. Using a specific antibody linked to fluorescein determine the location of the newly synthesized protein.

Interpretation: If the antibody stains the cytoplasm diffusely, the protein is synthesized but not targetted. If the outside of the nucleus is stained, producing a circular image, the protein binds but is not taken up. If the nucleus is stained, either uniformly or in patches, the protein contains the NLS and is transported.

General applicability: As an alternative to using the cDNA the protein itself can be placed in the cytoplasm. This can be achieved by microinjection, if large cells such as amphibian oocytes are used. Or the cells can be permeabilized, for example using digitonin or lysolecithin. In the latter case, it is necessary to demonstrate that the nucleus is not made permeable. Anti-DNA antibodies can be used for this purpose.

oxidase and enoyl oxidase all have the same C-terminal three amino acids, Ser-Lys-Leu (or SKL using the single letter nomenclature) [14]. The identification of the targetting sequence was determined experimentally.

- cDNA for luciferase containing C-terminal SKL was transfected into cultured cells. Luciferase was synthesized and concentrated in peroxisomes and detected using fluorescently labeled antibodies. *Therefore the experimental system is appropriate.*
- cDNA constructs of chloramphenicol acyltransferase (CAT) (a cytosolic enzyme) with C-terminal SKL were transfected into cultured cells. CAT was synthesized and concentrated in peroxisomes. *Therefore SKL is sufficient for targetting.*
- Capping of the SKL with a single amino acid prevents transport of the CAT into the peroxisomes. *Therefore SKL is specific and the location of the sequence is essential.*

Although SKL is the most common signal for peroxisome location, a second signal for peroxisome localization has been identified by similar experiments [15]. This consists of a nonapeptide at the amino terminal.

Peroxisomal membrane proteins are also synthesized in the cytosol and incorporated into the membrane post-translationally [16]. In contrast to most membrane proteins, some peroxisomal proteins are folded and oligomers assembled before translocation into the peroxisome. As yet the mechanisms have not been elucidated.

7.2.4 Targetting to mitochondria

Although mitochondria contain DNA, the majority of their proteins are coded for by nuclear DNA. Protein synthesis takes place in the cytosol and the proteins are transferred post-translationally into mitochondria. In principle, this transfer is similar to that into peroxisomes and nuclei. However, as the structure of the mitochondria is complex there are several potential destinations. These include the outer membrane, the inner membrane, the intramembrane space and the matrix . In recent years there has been considerable progress in identification of the proteins involved in the mitochondrial protein targetting and sorting [17,18] (*Figure 7.2*).

- Examination of the amino acid sequences or known membrane proteins demonstrated the existence of a targetting sequence for mitochondrial matrix proteins. This is not a consensus sequence but is rich in basic amino acids especially arginine. Structural analysis suggests that this sequence is an amphipathic α-helix.

- *In vitro* studies using isolated mitochondria (Box 7.4) have shown that the cytosolic chaperone hsp-70 is involved in maintaining the mitochondrial proteins in a form that is able to be translocated.
- A combination of genetic studies in which deletion mutants are identified and biochemical investigations of cell-free systems have identified an outer membrane and an inner membrane translocation complex (*Figure 7.2*) and have also shown that the membrane potential is necessary for protein sorting.
- After identification and purification of the proteins involved in translocation these can be reconstituted in liposomes and their mechanisms determined.

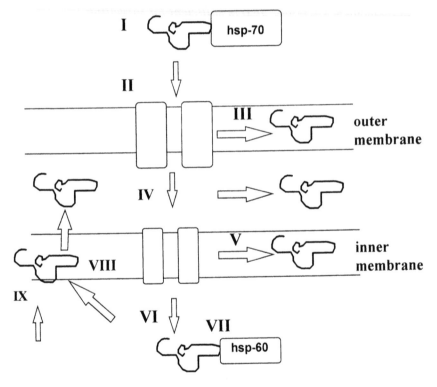

Figure 7.2. Molecular mechanisms involved in protein targetting in mitochondria. I, The newly synthesized proteins are kept unfolded by hsp-70 (or equivalent). This requires ATP. II, The protein docks on the outer membrane and moves through a pore complex. III, Some proteins remain membrane-associated, some move into the intramembrane space. IV, Proteins move from the intramembrane space through an inner membrane pore complex. V, Some remain membrane-bound and VI, some move into the matrix. VII, Hsp-60 (or equivalent) is involved in folding at this site. VIII, Some proteins move back into the inner membrane and the intramembrane space. IX, Proteins encoded for by the mitochondrial DNA are inserted into the inner membrane (see [17] for details).

Box 7.4

Aim: General approach to determine the mechanisms involved in sorting and translocation of mitochondrial proteins.

Protocol

Step 1. Isolate mitochondria (see Chapter 3).

Step 2. Over-express protein of interest in *E. coli* or prepare mRNA and translate this in a cell-free system (e.g. reticulocyte lysate).

Step 3. Incubate the protein and mitochondria. Check that the protein is translocated into the mitochondria by its resistance to added protease.

Step 4. Repeat the incubations, varying conditions to determine which co-factors are necessary.

Variations: The protein of interest can be cross-linked to putative proteins involved in translocation, or immunoprecipitated to determine which proteins are associated with, and at which stage during, translocation. Similar experiments can be carried out using purified outer membrane vesicles or inner membrane vesicles.

7.2.5 Targetting to the ER

All proteins destined for secretion, lysosomes or ER, Golgi or plasma membranes are directed initially to the ER. Proteins destined for nuclei, peroxisomes, mitochondria or chloroplasts are translocated post-translationally. In contrast, proteins destined for the ER and beyond are translocated co-translationally and this involves a complex molecular machinery (*Figure 7. 3*).

Evidence for the signal sequence includes (see Box 7.5):

- A cDNA construct having the putative signal from a secreted protein (e.g. β-lactamase) linked to a cytosolic protein (e.g. α-globin) transfected into cells produces a protein which is secreted.
- Deletion of the N-terminal signal from the cDNA of a secreted protein (e.g. β-lactamase) produces a cytosolic protein when transfected into cells.
- Deletion of the N-terminal signal from the cDNA of a plasma protein (e.g. influenza virus hemoglutinin) produces a cytosolic protein when transfected into cells.

The signal sequences of many proteins have been identified. Unlike the targetting sequences for peroxisomes the signal sequences are not the same in all proteins. These tend to be 15–50 amino acid residues

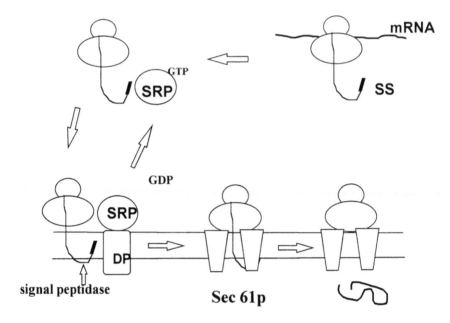

Figure 7.3. Molecular machinery involved in targetting proteins to the endoplasmic reticulum. The N-terminal which emerges from the large subunit of the ribosomes possesses the signal sequence (SS). The signal recognition particle (SRP) binds to this and arrests translation. GTP binds to SRP. The ribosome SRP complex binds to the docking protein (DP) or the SRP receptor. GTP is hydrolyzed and the GDP-DP releases SRP, which moves into the cytosol, and translation is reinitiated. The protein chain moves through a protein pore made up of the Sec61p complex. Note: a number of ribosomes read off the mRNA message simultaneously and form a polysome. All of the ribosomes in the polysome bind to the RER membrane.

at the N-terminus with a hydrophobic region between residues 8–18. The site for the signal peptidase also has a common structural motif consisting of a β-turn with small and neutral residues at −1 and −3 to the site of hydrolysis.

Evidence for the signal recognition particle

- The cell-free system described above containing mRNA, wheat-germ lysate and microsomal vesicles is used. When the microsomes are extracted with 0.5 M NaCl their ability to sequester newly synthesized protein is abolished.

Box 7.5

Aim: To determine whether a protein has a signal sequence.

Protocol

Step 1. Prepare mRNA from the cells of interest.

Step 2. Incubate mRNA with wheat-germ lysate which contains all of the factors necessary for synthesis of proteins including ribosomes. Add ER vesicles stripped of ribosomes (microsomes) (see Chapter 3).

Step 3. Pellet the microsomes by centrifugation.

Step 4. Separate the proteins by SDS–PAGE and detect the protein of interest by immunoblotting with a specific antibody (see Chapter 4).

Interpretation: If the protein of interest has a signal sequence it will be translocated to the endoplasmic reticulum vesicles. The signal sequence will be cleaved.

A   mRNA lacks signal sequence

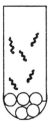

B   mRNA has signal sequence

immunoblot

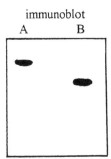

General application: This method can be used to demonstrate the presence of a signal sequence on any protein which can be specifically detected. Radiolabeled amino acids can be used in the system to aid in the detection of the protein of interest. In this case the protein can be detected on gels by autoradiography in addition to immunoblotting.

- Addition of the extract back to the system restores translocation of protein into microsomes.
- The signal recognition particle was isolated from the 0.5% M NaCl.

Evidence for the docking protein (or signal recognition particle receptor)

- Using the same system, treatment of the microsomes with elastase removed a further factor, which was identified as the docking protein.

7.2.6 Transport across the ER membrane

Proteins may be inserted into the RER co-translationally or post-translationally. The former mechanism, which is illustrated in *Figure 7.3* is generally found in mammalian cells, while both mechanisms occur in yeasts [19–21]. Secretory proteins move through the membrane while membrane proteins are partially translocated, so that the final topography is established (see Section 7.2.11 and Chapter 5). The current evidence suggests that newly synthesized proteins move across the RER through a protein-filled pore (Box 7.6).

7.2.7 Retention in the ER lumen

A number of proteins are resident in the lumen of the ER. These include enzymes which are involved in folding or modification of lumenal proteins. Secretory proteins may also be retained in the lumen of the RER. This occurs if the protein is not properly folded or requires another polypeptide for completion. In this case the secreted proteins are retained by binding to the resident proteins. One example is the heavy chain of immunoglobin associated with heavy chain binding protein (BiP). When the light chain is synthesized the heavy chain is released from BiP. The signal for retention of ER lumenal proteins was discovered by examining the amino acid sequence of known resident proteins, e.g. prolyl hydroxylase, protein disulfide isomerase and heavy chain binding protein (BiP). These all had the same C-terminal amino acids, Lys, Asp, Glu, Leu or KDEL [23]. Similar experiments to those described above for peroxisomal targets demonstrated that this sequence is sufficient for lumenal retention.

- Deletion of KDEL from the BiP cDNA, followed by transfection into cells, abolished endoplasmic reticulum retention and allowed the protein to be secreted.
- Addition of KDEL to the cDNA of a secreted protein, lysozyme, in a similar experiment caused retention in the endoplamic

reticulum. Changing the N-terminal to KDAS in place of KDEL caused secretion.

- KDEL inserted into the center of the protein did not result in endoplasmic reticulum retention.

Box 7.6

Aim: Demonstration that proteins move through a protein-lined pore.

Principle: Photoactivatable lysine residues are incorporated into the N-terminal of the nascent protein. These can be activated by light to cross-react with adjacent proteins.

Observation: The N-terminal attaches to two proteins. These are the putative protein pore.

Modification: The lysine residues were linked to nitrobenzyl hexanoic acid. This fluorescent molecule changes its emission intensity on movement from a polar to a nonpolar environment. This experiment shows that the pore is hydrophilic [22].

7.2.8 Transfer to the Golgi apparatus

In the ER the N-linked high mannose core oligosaccharides are added to newly synthesized proteins (see Chapter 2). These move to the Golgi and the mannoses are trimmed and sugars are added. The identification of the nature of the terminal sugar residues of the oligosaccharides thus provides a method for determining whether a protein has been transferred to the Golgi (see Box 7.7). For example, proteins resident in the ER membrane or lumen may be retained at that site after synthesis or may be retrieved from the Golgi. In the first case, proteins retained in the ER would have the high mannose oligosaccharides, while in the second they would have more complex sugars such as N-acetylglucosamine-galactose-sialic acid. Specific enzymes are used in these studies:

- Endoglycosidase-H (endo-H) cleaves mannose residues, but not if they are capped with other sugars. It therefore removes the high mannose oligosaccharides, i.e. added in the ER.
- N-Glycanase cleaves all N-linked oligosaccharides including those added in the ER and modified in the Golgi.

Box 7.7

Aim: To determine whether a protein has been modified in the Golgi.

Protocol

Step 1. Solubilize the membrane fraction or organelles under investigation.

Step 2. Treat aliquots with endoglycosidase-H (endo-H) and glycanase.

Step 3. Separate proteins by SDS–PAGE, electrotransfer on to nitrocellulose and detect protein of interest with specific antibody (see Chapter 4).

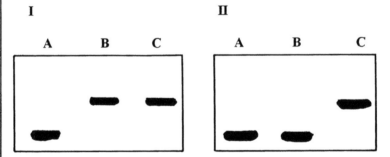

A = control; B = endo-H treated; C = glycanase. The removal of oligosaccharides alters the molecular weight of the protein. I, Result if protein has not been transferred to the Golgi; II, result of Golgi processing.

Applications: This method can be used to determine whether processing has taken place in the Golgi for either membrane or lumenal proteins.

7.2.9 Targetting to lysosomes

Lysosomes arise by budding from the Golgi membranes and contain a range of hydrolytic enzymes (see Chapter 2). Deficiencies in lysosomal enzymes are the molecular basis of a number of storage diseases. Investigation of one of the polysaccharide storage diseases, I cell disease, led to the identification of the molecular mechanism of targetting of enzymes to the lysosomes. The lysosomal enzymes are secreted in such patients and appear at high concentrations in the blood plasma. The enzymes in the plasma of patients were compared with those isolated from cultured fibroblasts of normal individuals. The mannose residues in the normal enzymes were found to be phosphorylated at the 6 position, while those in the patients were not phosphorylated, Mannose-6-phosphate is thus the target which sorts lysosomal enzymes. Phosphorylation takes place in the early Golgi and the sorting step involves receptors for mannose-6-phosphate in the late Golgi [24].

Phosphorylation of mannose is a two-step process. UDP-acetylglucosamine phosphate is transferred to the mannose residues in the first step. This is followed by removal of the UDP-acetylglucosamine.

*   The oligosaccharide alone is a poor substrate for phosphorylation. Therefore the specificity for phosphoryation lies in the lysosomal enzyme.
*   Heat-denatured lysosomal enzymes are not phosphorylated.
*   Fragments of lysosomal enzymes are not phosphorylated.

It is therefore considered that a structural motif is necessary for recognition and phosphorylation. This is termed a **signal patch**.

7.2.10 Transit of mannose-6-phosphatase receptors to and from the plasma membrane and Golgi membranes

Two forms of mannose-6-phosphatase receptors have been identified. One, the cation-independent form (CI-MPR), is located in both the Golgi and plasma membrane. The second, the cation-dependent form (CD-MPR), is only located in the Golgi. Approximately 80% of the lysosomal enzymes are diverted from the secretory pathway in the Golgi and targetted to the lysosomes. Some 20% escape sorting and are secreted. These are retrieved immediately by the plasma membrane receptors and taken up by endocytosis. Studies of the CI-MPR have provided information about the amino acid targetting sequences involved in movement of proteins between membrane compartments (Box 7.8) [25]. Such experiments can be applied to any membrane protein which has been cloned.

Box 7.8

Aim: To investigate the targetting sequences involved in targetting membrane proteins with CI-MPR as an example.

Protocol

Step 1. Prepare cDNA for CI-MPR in plasmid vector.

Step 2. Tranfect cells. Use fluorescent labeled antibody to locate the receptor to plasma membrane, endocytic vesicles or lysosomes.

Step 3. Repeat with CI-MRP truncated at COOH terminal or modified by site-directed mutagenesis.

Observations

- Removal of the C-terminal 40–89 residues (of the 169 residue cytosolic tail) stops sorting to lysosomes but not endocytosis. *Therefore the signal for endocytosis is in this sequence.*
- Removal of all but the 7–20 C-terminal cytosolic tail results in the receptors remaining at the cell surface. *Therefore the signal for endocytosis is in the cytosolic tail close to the membrane.*
- Site-directed mutagenesis of tyrosine 24 and 26 to alanine blocks endocytosis. *Therefore tyrosine at this site plays an important role – is it a site for phosphorylation or attachment to clathrin?*

7.2.11 *Establishing the topography and location of the membrane proteins of the ER, Golgi and plasma membrane*

Proteins which are targetted to the ER either move into the lumen or remain membrane-bound. The topography of the membrane proteins (*Figure 5.1*) is established co-translationally in the ER and is retained in those proteins which move on to the Golgi and plasma membrane. Although the mechanisms involved in establishing membrane topography have not been completely elucidated, there has been significant progress using both theoretical and experimental approaches [26].

- Analysis of the amino acid sequences in membrane proteins of known structure. As described in Chapter 5, the transmembrane sequences in membrane proteins are α-helices containing 20–25 predominantly hydrophobic residues. Clusters of positively charged residues tend to be located at the end of the α-helix in the part of the protein that is not translocated across the membrane.

- This has been confirmed by experiments in which positive residues are inserted into membrane proteins at critical locations. Using this approach it has been possible to reverse the direction of transmembrane regions.
- The signal sequence can act as a transmembrane α-helix. In this case the peptidase site is not present and the sequence is not cleaved.
- The protein chain is inserted into the membrane in a hair-pin shape. By combining noncleaved signals, distribution of positive residues and intramolecule hydrophobic regions, which form hair-pin loops, all types of membrane protein topographies can be modelled.

Membrane vesicles bud from the ER and move to the *cis*-Golgi. Vesicles also return from the *cis*-Golgi to the ER. This is collectively called the ***cis*-Golgi network**. Similarly vesicles move between the *trans*-Golgi and the plasma membrane and these form the ***trans*-Golgi network**. There are two possible mechanisms involved in determining the location of proteins within this membrane system:

- **retention** in the appropriate membrane, or
- **retrieval** back to the appropriate membrane by vesicular transport.

Both of these mechanisms have been implicated [26,27]. The main approaches to such studies are two-fold:

- isolation of mutants defective in protein sorting and identification of the mutated proteins, and
- transfection of cells with cDNA constructs having putative signals inserted or modified, followed by identification of changes in the way the protein is inserted into the membrane (see Chapter 5).

In the case of the ER-lumenal proteins with a C-terminal KDEL there is strong evidence for a retrieval mechanism involving a KDEL receptor in the *cis*-Golgi which returns the proteins to the ER. The ER membrane proteins also have identified retrieval signals.

- ER membrane proteins with a single transmembrane domain and a lumenal N-terminus have two critical lysines at the −3 and −4/5 positions relative to the carboxy-terminus.
- Proteins with a single transmembrane domain and the N-terminus on the cytosolic side have two critical arginines within the first five residues of the N-terminus. Incorporation of these residues into the protein structure is sufficient for ER membrane retrieval.
- It has also been shown that the transmembrane domains of Golgi membrane proteins are necessary for retention and that oligomerization may be important.

7.2.12 Covalently bound lipids in protein targetting

Acylation (see Chapter 5) has been shown to be involved in targetting of proteins to the plasma membrane. For example, the Rous sacoma virus protein P60 has covalently attached myristate on its N-terminal glycine residue. This protein is targetted to the cytosolic side of the plasma membrane. As the virus particle buds from the cell it takes with it a coat of plasma membrane containing the P60 protein.

* Deletion of the N-terminal seven amino acids prevents myristylation and also prevents targetting to the plasma membrane.
* Transfer of the N-terminal 14 residues from P60 to α-globulin results in myristoylation and targetting to the plasma membrane of the normally secreted α-globulin.
* The P21 ras protein is also targetted to the plasma membrane and is palmitoylated. Removal of the cysteine at 186 in the sequence CAAX prevents acylation and also prevents targetting to the cytosolic plasma membrane.

Phosphatidylinositol anchored proteins are targetted to the apical plasma membrane. The protein is linked to phosphatidylinositol in the RER through a specific recognition domain at the C-terminus (GPI anchor). A series of 'cut and paste' experiments, in which cDNA constructs were produced and transfected into cells followed by identification of the intracellular site of the newly synthesized proteins, have shown that the final location of the protein is determined by the GPI anchor.

* Addition of the GPI anchor converts human growth hormone from a secreted protein to a protein with an apical plasma membrane location.
* Addition of the GPI anchor converts VSV-G protein from a basolateral to an apical plasma membrane protein.
* The decay accelerating factor (DAF) is normally targetted to the apical plasma membane. Removal of the GPI anchor causes it to be secreted.

It should be noted that many plasma membrane proteins do not have covalently bound lipids. In these other targetting signals are involved.

7.1.13 Targetting in prokaryotes

Prokaryotic cells are structurally simple. The sites of proteins include the cytoplasm and the inner membrane. In the Gram-negative bacteria targetting to the periplasmic space and the outer membrane may also take place. The mechanisms of membrane targetting can be studied using similar techniques to those described for eukaryotes.

Table 7.2. Overview of signals involved in protein targetting

Signal	Target	Sequence or motif
No signal	Cytosol	
C-terminal SKL	Peroxisome	Specific and C-terminal
NLS	Nuclei	Non-specific sequence – high in basic amino acids
Presignal	Mitochondria	Non-specific amphipathic α-helix – high in Arg
Signal sequence	ER	Non-specific N-terminal – hydrophobic
Peptidase site	ER	β-turn; −1–3 rule applies
C-terminal KDEL	ER lumen	Specific and C-terminal
Signal patch	Lysosomal enzymes	Structural motif
Targetting sequences	ER membranes Golgi membranes Plasma membranes	Positive residues Transmembrane regions ??
Lipid anchors	Cytosolic plasma membrane Apical plasma membrane	Acylation GPI anchor

7.1.14 Overview

A variety of targetting signals are involved in sorting proteins (*Table 7.2*). These include structural or signal patches, specific sequences at specific sites in the protein, and nonspecific sequences at specific sites or at a range of sites in the proteins. The topography of membrane proteins is established by sequences in the protein molecule. The mechanisms involved in protein targetting include receptors involved in retrieval of proteins and also in retention of proteins at their characteristic locations.

References

1. **Schekman, R.** (1992) *Current Opin. Cell Biol.* **4:** 587–592.
2. **Armstrong, J., Pawell, E. and Pidoux, A.** (1992) in *Protein Targetting: A Practical Approach* (A.T. Magee and T. Wileman, eds), 87–109. IRL Press at Oxford University Press, Oxford.
3. **Bankaitis, V.A., Johnson, L.M. and Emr, S.D.** (1986) *Proc. Natl Acad.*

Sci. USA **83:** 9075–9579.

4. **Rothman, J.H. and Stevens, T.H.** (1986) *Cell* **47:** 1041–1051.
5. **Semenza, J.C., Hardwick, K.G., Dean, N. and Pelham, H.R.B.** (1990) *Cell* **61:** 1349–1357.
6. **Guthrie, C. and Fink, G.R.** *Methods Enzymol.* (1991) **194.**
7. **Tartakoff, A.M.** (1989) *Methods in Cell Biology*, Vol. 31, pp. 247–263. Academic Press, London.
8. **Hartl, F.U., Hiodan, R. and Langer, T.** (1994) *TIBS* **19:** 20–25.
9. **Georgopoulos, C.** (1992) *TIBS* **17:** 295–299.
10. **Landry, S.J. and Gierach, L.M.** (1991) *TIBS* **16:** 159–160.
11. **Pelham, H.R.B.** (1986) *Cell* **46:** 959–961.
12. **Butau, B., Hesterkamp, T. and Luirink, J.** (1996) *Trends Cell Biol.* **6:** 480–482.
13. **Gerace, L.** (1992) *Curr. Opin. Cell Biol.* **4:** 637–645.
14. **Gould, J.** (1989) *J. Cell Biol.* **108:** 1657–1684.
15. **Subramani, S.** (1996) *Curr. Opin. Cell Biol.* **8:** 513–518.
16. **McNew, J.A. and Goodman, J.M.** (1996) *TIBS* **21:** 54–58.
17. **Lill, R., Nargang, F.E. and Neupert, W.** (1996) *Curr. Opin. Cell Biol.* **8:** 505–512.
18. **Hohfeld, J. and Hartl, F.U.** (1994) *Curr. Opin. Cell Biol.* **6:** 499–509.
19. **Ng., D.T.W. and Walter, P.** (1994) *Curr. Opin. Cell Biol.* **6:** 510–516.
20. **Corsi, A.K. and Shekman, R.** (1996) *J. Biol. Chem.* **30:** 299–302.
21. **Rapoport, T.A., Rolls, M.M. and Jungnickel, B.** (1996) *Curr. Opin. Cell Biol.* **8:** 499–504.
22. **Johnson, A.E.** (1993) *TIBS* **18:** 456–458.
23. **Munro, S. and Pelhem, M.R.B.** (1987) *Cell* **48:** 899–907.
24. **Kornfeld, S.** (1987) *FASEB J.* **1:** 462–468.
25. **Hille-Rehfeld, A.** (1995) *Biochim. Biophys. Acta* **1241:** 177–194.
26. **von Heijne, G.** (1995) *Bioessays* **17:** 25–30.
27. **Luzio, J.P. and Banting, G.** (1993) *TIBS* **18:** 395–398.
28. **Nilsson, T. and Warren, G.** (1994) *Curr. Opin. Cell Biol.* **6:** 517–521.

8 Endocytosis and vesicular trafficking

In eukaryotic cells there is a flow of membrane from the rough endoplasmic reticulum to the plasma membrane during secretion (see Chapter 7) and a flow of membrane from the plasma membrane into the cytosol during endocytosis. This traffic is mediated by vesicles moving between the major organelles. These membrane vesicles thus contain a cargo, secreted molecules or ingested molecules, which must be delivered to the correct place, and the vehicle of transport, the membrane itself. Membrane traffic is tightly regulated to ensure the specific delivery of the cargo, while at the same time the membrane organelles maintain their unique protein and lipid composition. In general two mechanisms are involved in this complex sorting process – recycling of membrane components to achieve a steady state and retrieval mechanisms which correct sorting errors. Chapter 7 describes methods for studying protein targetting to the membrane organelles. This chapter will consider methods for studying receptors, endocytosis and the small transfer vesicles which shuttle between organelles.

8.1 Receptors and endocytosis

Endocytosis is the process by which regions of the plasma membrane invaginate, pinch off and transfer trapped extracellular materials into the cytosol. The process includes:

- *Pinocytosis* – the transfer of small membrane vesicles (< 150 nm diameter).
- *Phagocytosis* – the uptake of larger particles by vesicles which can be of a considerable size relative to the cell. Phagocytosis is a means of food intake in unicellular organisms and in multicellular organisms is only carried out by specialized cells (e.g. macrophages) while pinocytosis is carried out by most cell types.

- *Transcytosis* – vesicles form at one surface of polarized cells and move to fuse with another surface. This is a means by which materials cross epithelia, for example in endothelial cells of tightly coupled capillaries, or in the placenta, or in intestinal absorptive cells of neonates. Transcytosis is also a mechanism by which membrane proteins are transferred between regions of the plasma membrane, for example from the apical to the basolateral surface of epithelial cells.

8.1.1 Receptor–mediated endocytosis

This is a form of pinocytosis. Receptors are concentrated into specialized areas of the plasma membrane, which invaginate to form vesicles which move into the cytosol taking with them molecules bound to the receptors. Receptor-mediated endocytosis has a number of important roles in internalizing molecules which cannot cross the plasma membrane through transport proteins or channels.

- To take nutrients into cells, e.g. cholesterol which is taken up by the low density lipoprotein receptor, or iron which is taken up as a complex with transferrin.
- To inactivate polypeptide hormones by removing them from the circulation, e.g. removal of insulin by the insulin receptor.
- To facilitate the action of hormones, e.g. the insulin receptor which is a tyrosine kinase activated on binding insulin. Access of the hormone to the cytosolic proteins may be facilitated by endocytosis.
- To remove damaged proteins from the blood, e.g. the asialoglycoprotein receptor which removes glycoproteins that have lost their terminal oligosaccharide sialic acid.

8.1.2 Membrane receptors

These receptors bind a range of molecules all of which are called **ligands**. These include ions, small molecules (including many hormones) and polypeptides. A major role of receptors is in cell signaling and in most cases the ligand is not internalized but degraded extracellularly. Investigations of the properties of receptors have been aided by the use of analogs of the ligand, which may be synthesized in a radiolabeled form. These include:

- *Agonists* – which bind to the receptor and mimic the hormone action.
- *Antagonists* – which bind to the receptor and do not mimic the hormone. These are not always internalized or degraded so that

they provide useful tools for detemining the binding properties of receptors.

8.2 Investigation of events in endocytosis

Investigations of the kinetic events in endocytosis and receptor-mediated endocytosis fall broadly into three groups:

- Morphological investigation using the electron microscope.
- Biochemical investigation using cell fractionation, permeabilized cells and reconstituted systems.
- Genetic studies. As in investigations of the molecular events in secretion (see Chapter 7), genetic studies have also made considerable contributions to the identification of the molecules involved in endocytosis and vesicular trafficking. Convergence between the biochemical and the genetic approach has led to the identification of many of the gene products.

8.2.1 Electron microscopy

The morphological events in receptor-mediated endocytosis (see *Figure 8.1*) were initially elucidated through studies using the electron microscope. Although this method provides a static picture it is possible to obtain kinetic information by incubating cells with a molecule known to be internalized (ligand) and fixing the cells at a series of times for morphological examination. In this way intracellular transit of the ligand within the endocytic pathway can be followed. Ligands can be detected in a number of ways:

- By autoradiography, if a radiolabeled ligand is used.
- By immunocytochemistry, by binding a primary antibody to the ligand in thin sections followed by a secondary antibody conjugated to an electron-dense particle, for example colloidal gold.
- By immunocytochemistry in which the ligand or secondary antibody is coupled to an enzyme which has an electron-dense product, e.g. horseradish peroxidase.

Using this approach it was demonstrated that receptor-mediated endocytosis involves regions of the plasma membrane called coated pits, because the cytosolic surface has a fuzzy appearance in the electron microsocope. The coat is made up of clathrin adaptor protein complexes (*Figure 8.2*). Many receptors are concentrated in the coated pits through interaction between their cytosolic domains and the coat. Coated vesicles form when the pits invaginate and bud off the plasma

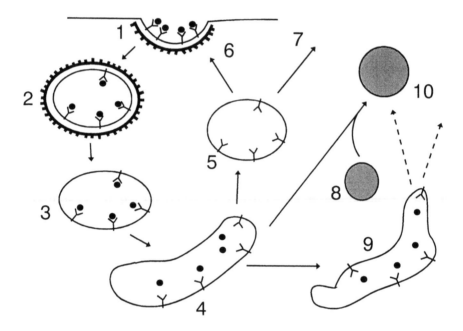

Figure 8.1. Pathways of receptor-mediated endocytosis. Schematic diagram showing some of the possible routes of intracellular processing of endosomes. 1: Receptor (Y)-bound ligand (●) endocytosed at coated pit to form coated vesicle (2) which forms a primary endosome (3). Primary endosomes carry receptor-bound ligand to tubular early endosome (4) which acts as a sorting station. A membrane-bound H^+-ATPase causes dissociation of ligand and receptor. Receptor may be returned in receptosome (5) to same plasma membrane domain (6) or to some other compartment (7) and ligand transferred to primary lysosome (8), forming secondary lysosome (10). One possible alternative is that the contents of the early endosome (4) may be transferred to or mature into a late endosome (9) before further processing of both ligand and receptor.

membrane. It is likely that coated pits are not static structures but rather the first step in the formation of coated vesicles arrested by fixation and processing for electron microscopy. In the cell, the coat is lost by disassembly and the clathrin and adaptor molecules return to the cell surface to form further coated pits. The vesicles fuse to form early endosomes, which mature to form late endosomes. The endosomes have a membrane-bound ATPase, which pumps protons into the vesicles. This causes a drop in pH, and the receptor–ligand complex separates. Further processing towards the late endosome stage results in recycling of the receptor back to the plasma membrane in vesicles and the ligand either enters the cell, for example iron or cholesterol, or is degraded by fusion of the vesicles with lysosomes (secondary lysosomes or late endosomes).

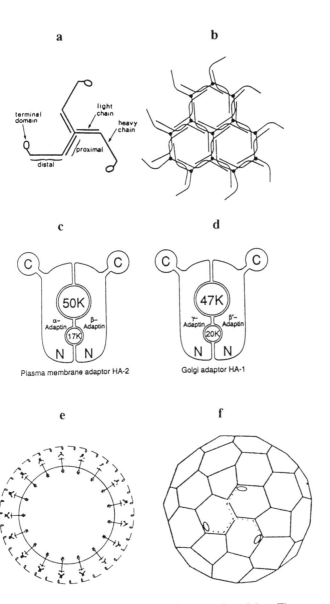

Figure 8.2. Components of clathrin-coated vesicles. The components of the clathrin triskelions are shown in (a). These are assembled to form five- and six-sided structures as shown in (b). The two forms of adaptors, HA-2 of the plasma membrane and HA-1 of the Golgi, are shown in (c) and (d), respectively. Adaptors bind to membrane proteins in the coated pits and recruit the clathrin triskelions to form a two-layered structure which buds off to form a coated vesicle, shown in cross-section in (e) and in surface view in (f). The arrangement of clathrin resembles that of buckminsterfullerine 'bucky balls'.

Box 8.1

Aim: To determine the equilibrium constant and the number of adrenergic receptors on adipocyte plasma membranes

Protocol

Step 1. Isolate adipocytes from adipose tissue by incubation with collagenase followed by low speed centrifugation to isolate the cells by flotation. Prepare plasma membrane vesicles [1].

Step 2. Incubate membrane with a range of concentrations of ^{3}H-dihydroaloprenolol (DHAP) (an adrenergic antagonist) at 37°C in buffer (50 mM Tris, pH 7.4, containing 10 mM $MgCl_2$).

Step 3. Isolate the plasma membranes by centrifugation or by filtration on to glass fiber discs. Wash the discs in the incubation buffer.

Step 4. Count the membranes or glass fiber discs in a scintillation counter. From the results calculate the amount of DHAP bound and the amount remaining unbound.

Step 5. Plot the moles of DHAP bound (abscissa) against the ratio of the bound DHAP divided by the free DHAP at equilibrium (Scatchard plot) [2]. The slope of the line equals $-K_a$, the affinity constant, and the intersection with the abscissa gives the number of binding sites (moles) in the membrane or cells used. The actual number of binding sites can be expressed in terms of membrane protein or per cell making use of Avogadro's number (1 mole = 6.022×10^{23} molecules).

Note: Care must be taken in the interpretation of the results as the Scatchard analysis is only valid if there is one class of binding site. If a curve rather than a straight line is produced this indicates that more than one binding site with different affinities are involved and the assay is invalid.

Controls: As DHAP is an antagonist its specificity can be determined by addition of the true hormone isoproteranol or propranolol. All specific binding will be blocked.

General applicability: This method can be used in principle for any cell type and any ligand. However, if internalization takes place this should be prevented by carrying out the incubation step at 0–4°C. In this case a control should include unlabeled hormone to determine nonspecific binding.

8.2.2 Biochemical studies

Determination of the number of membrane receptors and their affinity for the ligand. This can be investigated simply by incubating the membrane or cell under investigation with a radiolabeled ligand. For receptors not involved in receptor-mediated endocytosis it is preferable to use an antagonist which remains membrane-bound and is not degraded (Box 8.1). Nonspecific binding can be determined by addition of an excess of the true ligand or an agonist, which will compete for binding. If the receptor is involved in endocytosis the principles are the same; however, the incubation must be carried out at 0–4°C to prevent internalization.

Measurement of the rate of internalization and degradation of ligand.

If a radiolabeled ligand is available and the surface-bound ligand can be removed, for example by lowering the pH by addition of EDTA or by excess of unlabeled ligand, it is possible to measure the rate of internalization of the ligand and its degradation (Box 8.2). In an alternative method the surface exposed ^{125}I labelled ligand is available for immunoprecipitation while the internalized ligand is sequestered in vesicles and therefore not immunoprecipitated [3] (Box 8.3).

Measurement of the rate of internalization of a receptor and determination of the number of receptors.

This can be carried out by reversibly labeling the receptors with biotin. The amount of biotin bound to the membrane in the invaginated pits and in intracellular vesicles can be determined (Box 8.4).

Investigation of the fate of different ligands.

Using a relatively crude method of vesicle fractionation coupled with a method of selective density perturbation by selected ligands, the rate of internalization of ligands can be determined. Experiments can be designed to compare the fates of different ligands (Box 8.5).

Cell fractionation.

This can be used to separate the different compartments of the endocytic pathway on discontinuous and continuous gradients (see Chapter 3 and [5]). A major problem in such

---

Box 8.2

Aim: To measure the rate of ligand internalization and degradation

Protocol

Step 1. Incubate cells with radiolabeled ligand at low temperature to prevent internalization. Wash the cells to remove unbound ligand. Remove an aliquot of cells to determine the bound ligand at the beginning of the experiment.

Step 2. Incubate the cells at 37°C for a series of times. At each time point remove surface-bound ligand, for example by lowering the pH or using excess unlabeled ligand.

Step 3. Determine the level of radiolabel. The increase in label gives the rate of internalization.

Step 4. After removal of the surface-bound ligand, degradation of ^{125}I-tyrosine-labeled polypeptide ligands can be followed with time by precipitating the proteins with trichloroacetic acid. Degradation is indicated by conversion of insoluble radiolabel to soluble radiolabel and the rate of degradation calculated. To check that this involves lysosomal hydrolases, inhibitors of lysosomal enzymes should be added to the cell incubation medium.

studies is to identify subcellular fractions, as definitive markers have not been established. It is also unlikely that complete separation can be achieved. However, the endocytic pathway is functionally continuous and an analytical approach in which a spectrum of vesicles from the plasma membrane, coated vesicles, through to early and late endosomes are separated is an appropriate approach (Box 8.6).

8.3 Isolation of transport vesicles and investigation of the factors involved in vesicular transport

Conventional biochemical methods involving isolation of subcellular organelles or molecules disrupt the structure of cells and complex vesicular transport. It has therefore been necessary to develop novel systems to investigate the organization of cells and vesicular transfer between organelles. This approach, coupled with that of genetics, in which mutants deficient in transport systems have been produced, has led to considerable progress in indentification of transport vesicles and their functions.

8.3.1 Reconstitution of vesicular transport

The factors involved in vesicular tranport have been identified using a cell-free reconstituted system based on that developed by Rothman and co-workers [6] (*Figure 8.3*). This makes use of mutant cells deficient in the ability to catalyze the final steps in the synthesis of the oligosaccharide of secreted proteins, i.e. lacking N-acetylglucosamine transferase. These cells are infected with a virus, vesicular stomatitis virus (VSV). The viral G-glycoprotein is synthesized but the oligosaccharide is not completed. The Golgi stack is isolated from these cells (DONOR fraction) and mixed with Golgi stacks isolated from wild type uninfected cells, which contain functional N-acetylglucosamine transferase (ACCEPTOR fraction). If transfer of the viral protein takes place then some of this will move between the DONOR and ACCEPTOR membranes. If this is carried out in the presence of UDP ^{3}H-*N*-acetylglucosamine then the G-protein becomes radiolabeled. This can be immunoprecipitated and counted. The radiolabel incorporated into the G-protein is proportional to the transfer between the Golgi stacks, as such transfer is needed for the exposure of the G-protein to *N*-acetylglucosamine transferase. This system has been used to determine which factors are necessary for vesicular transport and for the isolation of the transport vesicles which form *in vitro*.

Box 8.3

Determination of the distribution of transferrin receptors between the cell surface and the internal vesicles

Protocol

Step 1. Label cells with ^{125}I-transferrin.

Step 2. Rupture cells.

Step 3. Add anti-transferrin. This binds to transferrin on the cell surface but not to that internalized in vesicles.

Step 4. Add a large excess of unlabeled transferrin to bind the antibody.

Step 5. Detergent-solubilize cells, e.g. with Nonidet, and precipitate the transferrin–anti-transferrin antibody complexes with Protein A or Protein G bound to beads.

Step 6. The radiolabel in the immunoprecipitate is the surface receptor.

Step 7. Carry out the same procedure but add the ^{125}I-transferrin and antibody after solubilization. In this case the radioactivity in the immunoprecipitate is a measure of total receptor.

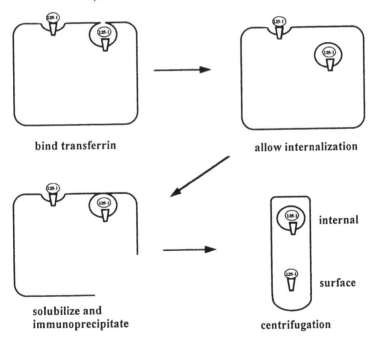

bind transferrin   allow internalization

solubilize and
immunoprecipitate   centrifugation

internal

surface

Box 8.4

Aim: To determine the number of transferrin receptors in cells and the rate of their internalization

Principle: Biotin forms an S–S bond with free SH groups (cysteine residues) in the extracellular domain of the receptor under the oxidizing conditions of the incubation medium. Some of the biotin will be internalized. Addition of a reducing agent causes available biotin to be removed from the receptor (i.e. that at the cell surface or in invaginations). The biotin bound to the receptor in intracellular vesicles is determined by immunoprecipitation of the transferrin receptor, separation of the proteins by SDS–PAGE and indentification of the receptor by avidin binding. The amount of avidin, and hence the amount of biotin, can be determined using radiolabeled avidin or an anti-avidin antibody by quantitative immunoblotting.

By removal of samples at different times after labeling the receptor with biotin the rate of internalization can be determined. Using a pulse-chase procedure the rate of return of the receptors to the plasma membrane can be determined.

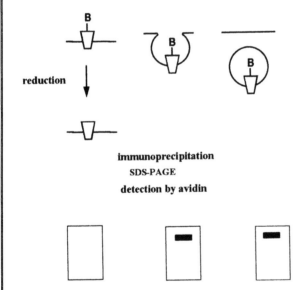

The amount of biotin at the surface of the plasma membrane can be determined by binding biotin to the cells followed by avidin, which is excluded from the invaginations, followed by assay of the avidin immunoprecipitated as above.

General applicability: This method can be used for any receptor which contains cysteine residues which will form S–S bonds with biotin.

Box 8.5

Aim: To determine whether two receptors are in the same coated pit and vesicle in HepG^2 cells [4]

Principle: Cells are incubated with two different ligands labeled with different isotopes. One ligand is conjugated to horseradish peroxidase and has a dense product when provided with the appropriate substrate. The cells are disrupted to release coated vesicles, endosomes and secondary lysosomes. The heavy and light vesicles are separated by density gradient centrifugation (see Chapter 3) and the ligands in the vesicles identified by counting.

Step 1. Incubate Hep-G2 cells with transferrin labeled with ^{131}I and asialoglycoprotein labelled with ^{125}I and conjugated to horseradish peroxidase.

Step 2. At a series of times, rupture the cells to release the vesicles.

Step 3. Incubate the mixture with 3,3'-diaminobenzidine and hydrogen peroxide to allow a dense product to accumulate in those vesicles containing the asialoglycoprotein.

Step 4. Separate the dense vesicles from the lighter vesicles by centrifugation on a suitable gradient (Chapter 3).

Step 5. Assay [125]I (marker for transferrrin) and [131]I (marker for asialoglycoprotein).

Observations

- 1–2 min after incubation both ligands are in the dense vesicles.
- 2–10 min after incubation only asialoglycoprotein is in the heavy vesicles, indicating that sorting has taken place and transferrin is returned to the cell surface.
- At later times asialolglycoprotein is degraded.

General applicability: This method can be used to follow the kinetics of transfer of a single ligand through the endocytic pathway and can be used to identify the morphological components and kinetic relationships between the different vesicles.

8.3.2 Semi-intact cell systems

In this system, cultured cell monolayers are cooled to 20°C to arrest protein transport. The cells are then permeablized, either by using a detergent such as saponin which removes plasma membrane cholesterol, or by allowing a nitrocellulose membrane to adhere to the monolayer surface (*Figure 8.4*). The nitrocellulose is removed, taking off the surface membrane. The cytosol is lost using either method. When the cells are warmed, transport is restarted. The factors necessary for transport can be investigated by addition of cytosol or its components and the transport vesicles can be isolated. This approach is usually coupled with infection of the cells, for example by VSV, when the presence of the viral G-protein facilitates isolation and identification of transfer vesicles, using an affinity column against the cytosolic tail of the G-protein.

Box 8.6

Investigation of the transit of ligands through the endocytic pathway

Step 1. Incubate cells with a radiolabeled ligand.

Step 2. At a series of times prepare a crude microsomal or large vesicle fraction and separate this on a continuous gradient, for example on a self-generating iodixanol or Percoll gradient, or a preformed sucrose gradient (see Chapter 3).

Step 3. Collect the gradient in a series of fractions.

Step 4. Determine the distribution of the labeled ligand on the gradient at each time. This may be expressed as % recovery of radiolabel in each fraction or as d.p.m. in each fraction.

Step 5. Use marker enzymes or membrane components to identify the subcellular fractions (see Chapter 3).

Interpretation: The plasma membrane is the first organelle radiolabeled followed by coated vesicles, early endosomes and late endosomes. The rate of transit can be followed and the effect of reagents known to perturb specific cell functions investigated (see *Table 8.1*).

General applicability: If more refined methods for preparation of defined fractions are available, these can be used to follow the transit of a specific ligand or membrane protein.

Table 8.1. Examples of reagents which perturb intracellular transport and ligand processing

Reagent	Site of action
Colchicine	Microtubules
Nocodazole	Microtubules
Brefeldin A	Coat proteins of vesicles between ER and Golgi
Monensin	Acidic organelles – lysosomes
Chloroquin	Acidic organelles – lysosomes
Vanadate	Phosphorylation

8.4 Transfer vesicles

There has been considerable progress in recent years in the identification of the different transfer vesicles involved in membrane

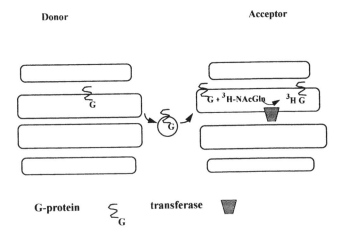

Figure 8.3. Cell-free system for investigation of membrane transfer. Incorporation of ³H into immunoprecipitated G-proteins is an indicator of vesicular transport (see text for details).

traffic and in the determination of the molecules involved in this complex process. A brief outline of the vesicles and their components is described below. For further information the reader is referred to the many recent reviews [3,6–25].

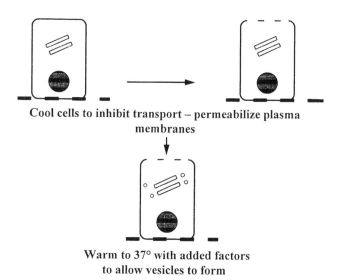

Figure 8.4. Semi-intact cell system for investigation of membrane vesicular transport (see text for details).

8.4.1 Transfer from the plasma membrane to the endosomes

These vesicles have a coat made up of adaptor 2 complexes which link vesicle proteins to clathrin (see *Figure 8.3*) [8–11]. Dynamin is involved in the fusing of the plasma membrane to form a coated vesicle [10]. The adaptor complexes are recruited by plasma membrane proteins and recognize a specific motif on the cytosolic domain of the membrane protein, tyrosine-X-X-hydrophobic residue (Y-X-X-Z, where X is any amino acid) [19]. This recognition sequence was established by mutagenesis and its association with adaptor 2 by the use of the yeast two-hybrid system.

8.4.2 Transfer from the Golgi to lysosomes

These vesicles have a coat made up of adpator 1 complexes which link the vesicle proteins to clathrin (*Figure 8.3*) [11,12].

8.4.3 Golgi to endoplasmic reticulum (retrograde transport)

The retrieval of proteins from the Golgi back to the endoplasmic reticulum is carried out by COP1 vesicles [13–16, 22, 23]. The budding process is initiated by ARF, a GTPase, which moves from the cytosol to bind to the appropriate region of the membrane. Coatamer proteins then bind to ARF and form a vesicle. The coatamer proteins include seven subunits, which are in the cytosol until activated. The ER membrane proteins have a motif lysine-lysine-X-X (K-K-X-X), where X is any amino acid, recognized by components of the COP1 [18]. If vesicular transport is blocked, for example by using a nonhydrolyzable analog of GTP (GTPγS), then COP1 vesicles accumulate. This prevents the return of membrane proteins to the ER and blocks forward transport from ER to Golgi.

8.4.4 Endoplasmic reticulum to Golgi (anterograde transport)

Forward transport of proteins from the ER to the Golgi involves COP2 vesicles [22,23]. The components of these were identified by genetic methods and probably function in a similar way to the formation of COP1 vesicles. The GTPase, Sar 1p, is an ARF homolog and the coatamer proteins include Sec 13p and Sec 23p. The isolated proteins have been shown to reconstitute COP2 vesicles *in vitro*.

8.4.5 Recognition in vesicular transfer

It is essential that the transfer of vesicles is carried out with high fidelity. This is achieved by the presence of specific membrane receptors, v-SNARES in the vesicle and t-SNARES in the target membrane [6]. SNAPS and NSF allow v-SNARES and t-SNARES to bind in stable specific complexes [24,25]. RabGTPases catalyze the accurate association of the vesicles and their target membranes [17].

References

1. **Bennett, V. and Cuatrecasas, P.** (1973) *Biochim. Biophys. Acta* **311**: 362–380.
2. **Scatchard, G.** (1966) *Ann. N.Y. Acad. Sci.* **51**: 660–720.
3. **Warren, G., Woodman, P., Pypaert, M. and Smythe, E.** (1988) *TIBS* **13**: 462–465.
4. **Stoorrvogel, W., Geuze, H.J. and Strous, G.J.** (1987) *J. Cell. Biol.* **104**: 1261–1268.
5. **Gjoen, T., Berg, T.O. and Berg, T.** (1997) in *Subcellular Fractionation, A Practical Approach* (J.M. Graham and D. Rickwood, eds), pp. 169–200. IRL Press at Oxford University Press, Oxford.
6. **Rothman, J.E.** (1994) *Nature* **372**: 55–63.
7. **Anderson, R.B.W.** (1993) *Trends Cell Biol.* **3**: 177–179.
8. **Pease, B.M.F. and Robinson, M.S.** (1990) *Ann. Rev. Cell Biol.* **6**: 150–171.
9. **Robinson, M.S.** (1997) *Trends Cell Biol.* **7**: 99–102.
10. **Robinson, M.S.** (1994) *Curr. Opin. Cell Biol.* **6**: 538–544.
11. **Ludwig, T., Le Borgne, R. and Hoflack, B.** (1995) *Trends Cell Biol.* **5**: 202–208.
12. **Hunziker, W. and Geuze, H.J.** (1995) *Bioessays* **18**: 369–379.
13. **Kreis, T.E. and Pepperkok, R.** (1994) *Curr. Opin. Cell Biol.* **6**: 533–537.
14. **Bednarek, S.Y., Orci, L. and Schekman, R.** (1996) *Trends Cell Biol.* **6**: 468–473.
15. **Donaldson, J.G. and Klausner, R.D.** (1994) *Curr. Opin. Cell Biol.* **6**: 527–532.
16. **Boman, A.L. and Kahn, R.A.** (1995) *TIBS* **20**: 147–150.
17. **Pfeffer, S.R.** (1994) *Curr. Opin. Cell Biol.* **6**: 522–526.
18. **Sandoval, I.V. and Bakke, O.** (1994) *Trends Cell Biol.* **4**: 292–297.
19. **Marks, M.S., Ohno, H., Kirchhausen, T. and Bonifacino, J.S.** (1997) *Trends Cell Biol.* **7**: 124–128.
20. **Seaman, M.N.J., Burd, C.G. and Emr, S.D.** (1996) *Curr. Opin. Cell Biol.* **8**: 549–556.
21. **Gruenberg, J. and Maxfield, F.R.** (1995) *Curr. Opin. Cell Biol.* **7**: 552–563.

22. **Aridor, M. and Balch, W.E.** (1996) *Trends Cell Biol.* **6:** 315–320.
23. **Salama, N.R. and Schekman, R.W.** (1995) *Curr. Opin. Cell Biol.* **7:** 536–543.
24. **Sollner, T.** (1995) *FEBS Letts* **369:** 80–83.
25. **Whiteheart, S.W. and Kubalek, E.W.** (1995) *Trends Cell Biol.* **5:** 64–68.

Index